Methods in Cancer Stem Cell Biology

Said M. Afify • Masaharu Seno

Methods in Cancer Stem Cell Biology

Springer

Said M. Afify
Department of Oncology
Lombardi Comprehensive Cancer
Georgetown University
Washington, DC, USA

Division of Biochemistry
Chemistry Department
Faculty of Science
Menoufia University
Shebin El Koum, Menoufia, Egypt

Graduate School of Interdisciplinary
Science and Engineering in Health Systems
Okayama University
Okayama, Japan

Masaharu Seno
Graduate School of Interdisciplinary
Science and Engineering in Health Systems
Okayama University
Okayama, Japan

ISBN 978-981-99-1330-5 ISBN 978-981-99-1331-2 (eBook)
https://doi.org/10.1007/978-981-99-1331-2

This Springer imprint is published by the registered company Springer Nature Singapore Pte Ltd.
The registered company address is: 152 Beach Road, #21-01/04 Gateway East, Singapore 189721, Singapore

This book is dedicated to my parents whose love and devotion has illuminated my path and made each step joyful.

—From Said M. Afify

This book is dedicated to Prof. Katsusaburo Yamagiwa and Prof. Satimaru Seno.

—From Masaharu Seno

Preface

During the last century, cancer has been considered as a genetic disease. It is true that cancer induction has been demonstrated by the introduction of mutations and oncogenes into normal cells. And this has been so-called transformation. On the other hand, the initiation of cancer without introduction of genetic alterations has hardly been studied. Probably, Prof. Katsusaburo Yamagiwa was the first to work on cancer initiation. His thought of cancer has become more prevalent than ever due to the constitutive exposure to carcinogenic factors in the environment such as chronic inflammation. Although embryology and developmental biology are closely approaching this issue, fewer people appear to succeed his laudable work. In this context, the development of induced pluripotent stem cells (PSCs) was really epoch making because we could start studying cancer initiation by producing cancer stem cells (CSCs). Many different studies have been published on our unique approach during the decade to convert stem cells including PSCs into CSCs without introducing mutations or foreign gene. These studies not only characterize the cancer stem cells but also define the cancer-inducing niche that alters the state from normal to diseased. Given that cancer causes significant death rates worldwide despite a century of research, the use of a suitable technique in cancer biology with a rather novel concept appears essential for both basic research and clinical application. Although there are many books attempting to evaluate methods in cancer biology on the market, none of them describes the use of stem cells and cancer stem cell generation in the inflammatory microenvironment as the cancer-inducing niche using stem cells such as PSCs. The aim of this book is to provide step-by-step techniques and bench manuals for the study to develop CSCs from normal stem cells with different applications. The essential techniques on stem cell and CSC biology are provided in this book. This is in part related to the field of microenvironments and cancer initiation, asking how stem cells play a role in them. The editors believe that our understanding of the cell and molecular biology of cancer has greatly been changed through the experiments with the CSCs derived from PSC. Our goal is to share these experiences with scientists in the world providing this book as a handbook to deepen cancer research. The editors hope this book becomes a good guidance and help for readers.

The book consists of 20 chapters that include methods to generate PSCs, maintain normal stem cells, convert normal stem cells to CSCs, enrich, isolate, and evaluate CSCs. Throughout this book, we describe how cancer stem cells can be generated from normal stem cells as the context. Illustrations may help understand how the method of CSC generation can be applied as an essential method to assess the carcinogenic potential of various non-mutagenic compounds.

Observations, hypotheses, and practices that have shaped modern medicine during a couple of past decades are discussed in Chap. 1. Cell culture method is introduced in Chap. 2 followed by a description of the general techniques used in many cell culture facilities. Chapters 3–5 cover the embryonic stem cell culture, the techniques to reprogram normal cells into PSCs as well as the conditions to maintain and monitor human PSC. A variety of methods are described in Chaps. 6–8 for identification, enrichment, and isolation of CSCs, which will facilitate the establishment of CSC lines from malignant tumors for the future work in vitro. Chapter 9 outlines the process of converting stem cells into CSCs, and Chap. 10 exploits this process to assess the carcinogenic potential of chemicals. Especially, this method is available for assessing the effects of non-mutagenic chemicals while the methods to assess mutagens as carcinogens are more popular. In Chaps. 11–17, various methods are described to evaluate the characteristics of CSCs in vitro and in vivo such as the abilities of self-renewal, differentiation, and tumorigenicity as well as the metastatic potential. In Chaps. 18–20, possible targeting therapies against CSCs are described. The studies on the development of CSCs from normal PSCs, which will help design new therapies applicable to treat CSCs, should open a new page of cancer research. We tried to make the book easy to follow, so that researchers seeking new techniques related to CSCs should feel easy to introduce the book advantageously to their laboratories. This book will also be useful for not only graduate students but also scientists, technicians, and physicians working in academic, hospital, or pharmaceutical settings.

Okayama, Japan

Said M. Afify
Masaharu Seno

Acknowledgments

It is a long-term project to write this book, and I thank the many people who have contributed, consciously or not, to this successful achievement. First of all, I would like to thank my God for giving me this opportunity and to complete my work.

So, I greatly thank all lab members who joined our lab from 2012 to this day. Without their great contribution, this project will not be accomplished now.

In writing this book, I am very grateful to **Dr. Anna Sanchez Calle, Dr. Maram Zahara, Dr. Hend Nawara, Dr. Ghmkin Hassan, Dr. Amira Osman, and Dr. Samah Elghlban** for being my inspiration. I gratefully acknowledge the funding received towards my PhD from **MEXT** (Ministry of Education, Culture, Sports, Science and Technology) Japan. Also, I cannot forget to acknowledge the funding received towards my postdoctoral position from **JSPS** (Japan Society for the Promotion of Science).

I would like to thank my professors and colleagues from the Faculty of Science, Menoufia University, and the Department of Biochemistry in the Faculty of Science, Cairo University, for their support and encouragement during my studies.

For no reason, and for all reasons, I would like to thank my wife **Dr. Hend Nawara.** You are not just my partner, you are my lover. You are not just my companion, you are my inspiration. You are not just my wife, you are my life.

As a final note, I must mention my **parents** and my **daughters**, **Remas** and **Rokaia**, who have continually and unconditionally supported me throughout my life.

SAID M AFIFY, PhD

I acknowledge the professors and colleagues in the Laboratory of Nano-Biotechnology, Graduate School of Interdisciplinary Science and Engineering in Health Systems and Graduate School of Natural Science and Technology, Okayama University, for their support and encouragement.

I appreciate the most important and impressive contribution of Dr. Ling Chen who is the first to work on our CSC study. We also appreciate all the descendants who succeeded in their PhD study establishing and using CSCs raising their names. They are Dr. Shuichi Matsuda, Dr. Yan Ting, Dr. Marta Prieto Vila, Dr. Anna Sanchez Calle, Dr. Neha Nair, Dr. Aun Ko Ko Oo, Dr. Md Jahangir Alam, Dr. Said

Afify, Dr. Juan Du, Dr. Hend Nawara, Dr. Hager Mansour, Dr. Ghmkin Hassan, Dr. Hagar Ali Abu Quora, Dr. Kazuki Kumon, Dr. Mona Sheta, Dr. Sadia Monzur, Dr. Hideki Minematsu, and Dr. Yanning Xu. Further, I thank pathologists who have supported our in vivo studies. They are Prof. Yoshiaki Iwasaki, Prof. Toshiaki Ohara, Prof. Xiaoying Fu, and Dr. Amira Osman.

Masaharu Seno, PhD, Professor Emeritus

Contents

Chapter 1
On the Origin of Cancer

Abstract Cancer is a disease with a long history from prehistoric times that could be traced back to the era of dinosaurs. In humans, cancer becomes prevalent much more than ever due to the extension of average lifespan and the increase of exposure to carcinogenic factors in the environments during the last decades. Seven Egyptian papyri, which were the oldest in human history to introduce Egyptian medical practice, were found and deciphered in the late nineteenth century. "Edwin Smith" and "George Ebers" were the two papyri that contained the descriptions of cancer. They were estimated to be written between around 3000 to 1600 BC. Hippocrates (460–370 BC), the "Father of Medicine," used the term "karkinos," which means a crab in Greek, to describe the disease since a tumor with metastasis along with lymph nodes appeared like crab finger projections from the breast. And probably breast cancer cases were the easiest ones to find and diagnose at a glance.

Since the beginning of cancer study, many theories about cancer have been proposed including humoral theory, lymph theory, blastema theory, chronic irritation theory, viral oncogenesis, chemical carcinogenesis, mutation theory, tissue organization field theory (TOFT), and cancer stem cell theory. In the current chapter, we highlight old observations, hypotheses, and practices in cancer research shaped as modern medicine during a couple of past decades with excerpts.

Keywords Humoral theory · Lymph theory · Blastema theory · Chronic irritation theory · Viral & Chemical carcinogens · Stem cell

1.1 Background

There was a time when humans began to observe diseases in the history of humanity. They also attempted to know the origin and cause of cancer as we are currently doing. Through the long history of thousands of years, different types of cancers were observed and a number of different therapies such as herbs, salt mixtures, surgery, and so on were applied. You can find the earliest documents on cancer described in ancient Egypt on several papyri. An ancient Egyptian medical text, so-called Edwin Smith Papyrus (Fig. 1.1), written around 1700 BC, is thought to describe medicine dating back to approximately 3000 BC. A technique to remove

S. M. Afify, M. Seno, *Methods in Cancer Stem Cell Biology*,
https://doi.org/10.1007/978-981-99-1331-2_1

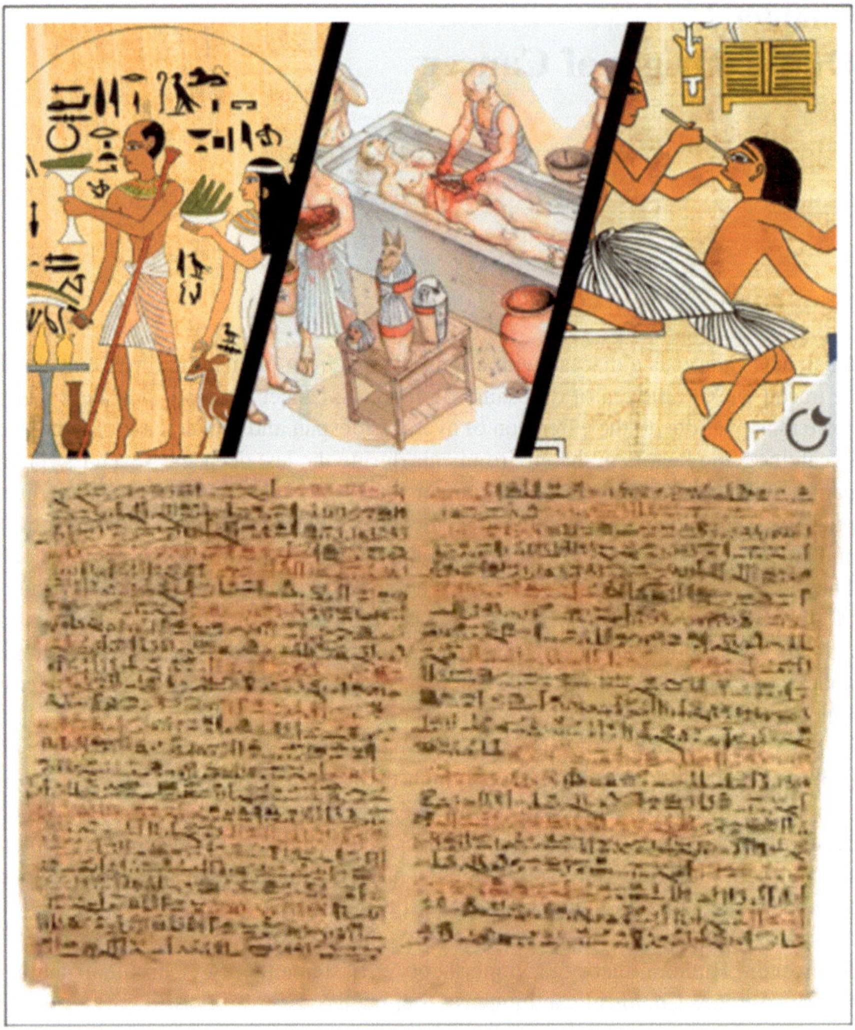

Fig. 1.1 The Edwin Smith Papyrus was written about 3000 BC

tumors is found in it with eight cases of breast tumors, which were cauterized with a fire drill. We can also read that they knew tumors could not be treated when they became cold, bulging, or spread over the breast (Breasted 1930). The Ebers Papyrus is another Egyptian medical text written in 1500 BC, in which soft tissue tumors were described as fatty tissue cancer found in the skin, uterus, stomach, and rectum (Ebbell 1937). "Egyptian ointment" is the ancient treatment of cancer with knives, cautery, salts, and arsenic paste, which was commonly applied until the nineteenth century (Hajdu 2004).

The oldest documented cases of cancer in humans are often found in Egyptian and Peruvian mummies made latest in 1500 BC. In 2700 years ago, a 40- to 50-year-old Scythian king living in the steppes of southern Siberia was found to have disseminated cancer.

The medicine from the ancient era of not only Egyptian but also Greek, Persian, Indian, and Chinese cultures has led to various and helpful insights and important scholarly achievements in cancer treatment combining the study of the human body with the study of therapeutics.

All these efforts made to understand cancer disease during this long time have resulted in many theories until now, so here we will try to cover most of them to provide an understanding of the history of cancer that will direct our understanding of its future.

1.2 Humoral Theory

A significant progress was introduced by the Greeks around 400 BC, around 100 years after Hippocrates who developed a better understanding of cancer by referring to the disease as "Karakinos" named after the moveable crab that adheres to its surroundings with his claws. A Roman encyclopedist and physician whose name was Aulus Cornelius Celsus (25 BC–50 AD) later translated the word "Karkinos" to "cancer," which means "crab" in Latin, in his book, De Medicina (Fig. 1.2).

In historical accounts, a Greek named Galenus lived between 129 and 216 AD and was credited with introducing humoral theory under the name of Hippocrates (Hajdu 2004, 2011; Castiglioni 1931; Major 1954). Humoral theory became one of the central principles in Western medicine from antiquity to the nineteenth century, until when anatomical pathology began. The word "humoral" derives from the word "humor," which means "fluid" in this context. He described that the body contained four humors (body fluids): blood, phlegm, yellow bile, and black bile. According to this theory, a person was healthy when the humors were balanced. It was also believed that an excess or a shortage of any single humor caused disease. Galenus described cancer in detail in his text "De Tumoribus Praeter Naturam." He believed that thick black bile caused ulcerated and incurable cancer, and thin yellow bile caused non-ulcerated and curable cancer. Galenus recommended no treatment for cancer, unless the cancer involved the skin, in which case surgery would be chosen as the treatment. He thought cancer was a disease that should be treated with purgatives to decrease the accumulation of black bile.

1.3 Lymph Theory

Andreas Vesalius (1514–1564), a great Flanders anatomist of the sixteenth century, readily understood the human body during his direct observations. Although he did not find any black bile, he further found lymph and described how it flowed through

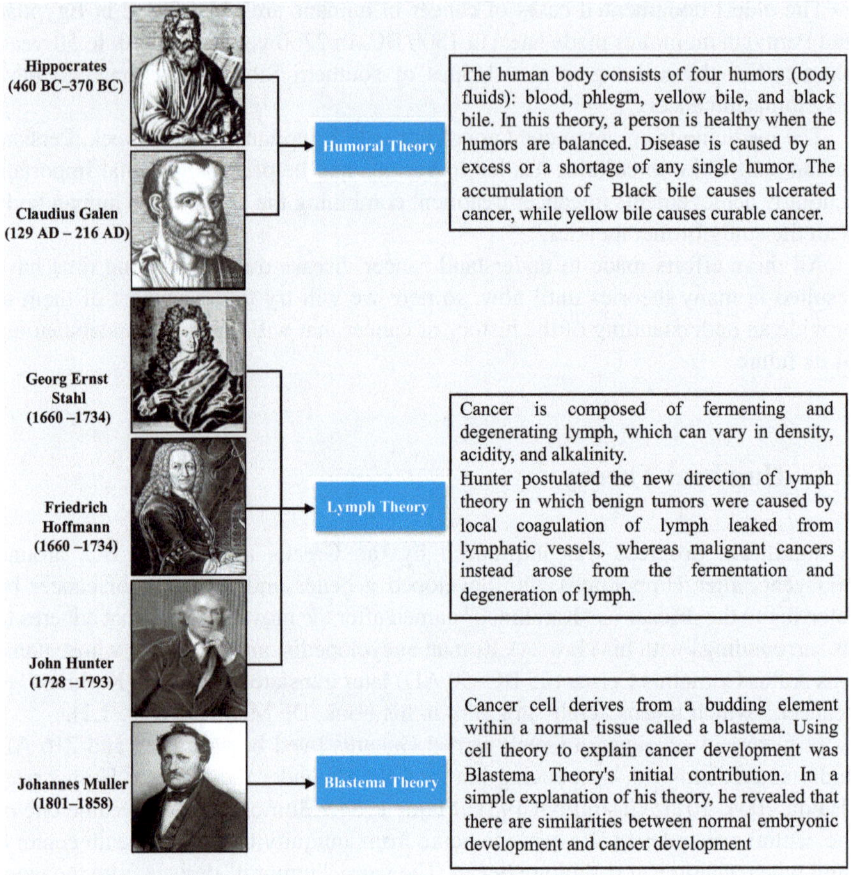

Fig. 1.2 Classical concepts of cancer up to the middle of the nineteenth century from the humoral to the blastema theories

channels. Due to these experiences, Vesalius began to doubt Galenus's theory (Diamandopoulos 1996; Kardinal and Yarbro 1979; Javier and Butel 2008). However, he hushed up his observations burying them in his unpublished papers.

It was two hundred years after Vesalius when Matthew Baillie (1761–1823), a British physiologist and anatomist, described the appearance of many tumor specimens, none of which contained black bile. This evidence not only killed the black bile theory but also revived the surgery removing tumors, which became the major cancer treatments in the following centuries.

The concept of cancer made a great stride with the rediscovery of lymph by Gasparo Aselli (1581–1625) in 1622 (Suy et al. 2016) and the demonstration of blood circulation by William Harvey (1578–1657) in 1626 (Ribatti 2009). Based on this information, Georg Ernst Stahl (1660–1734) and Friedrich Hoffmann

(1660–1742) developed the Lymph theory. They thought that another body fluid, lymph, was responsible for the development of cancer, replacing the humoral theory while homeostasis was maintained by the proper movement of fluids, mainly blood and lymph, through solid body parts. According to the lymph theory of cancer, Stahl and Hoffman proposed that cancer was composed of fermenting and degenerating lymph, which varies in density, acidity, and alkalinity.

A new direction was given in the lymph theory by John Hunter (1728–1793), a professor of anatomy and surgery in London. Hunter characterized lymph by coagulation, in which blood was exuded from vessels and became coagulated while the classical lymph was characterized by degeneration. Hunter's theory defines that tumors arise as a result of the activity of the organism itself. They are similar to normal tissue. They namely live, grow, and change as the normal tissues in the body do.

1.4 Blastema Theory

Johannes Müller (1801–1858), a German pathologist, demonstrated in 1838 that cancer was made up of cells but not lymph (Johannes Müller, 1938). He believed cancer cells were not histologically normal. Based on his assertions, a cancer cell derives from a budding element so-called blastema within a normal tissue (Fig. 1.1). The blastema theory importantly contributed to explain cancer development by introducing the concept of a cell. He revealed that there were similarities between the development of embryos and cancer.

1.5 Chronic Irritation Theory

Rudolph Virchow (1822–1902) hypothesized that cancer arose from severe irritation that leads to inflammation and excessive cellular proliferation resulting in cancer. This theory became known as "chronic irritation theory" (Fig. 1.3). Virchow, a student of Johannes Mueller, was one of the first who accepted the concept of cell division and believed that connective tissue could support the development of all malignant tumors.

In 1863, Virchow described that cancer spread like a liquid following chronic irritation. He proposed some classes of irritants by their tissue injuries and subsequent inflammation enhancing cell proliferation. According to him, the origins of tumors were primarily due to three factors:

1. The local situation.
2. A predisposition to disease based on the individual's constitution.
3. Dyscrasia, a condition of the body liquids.

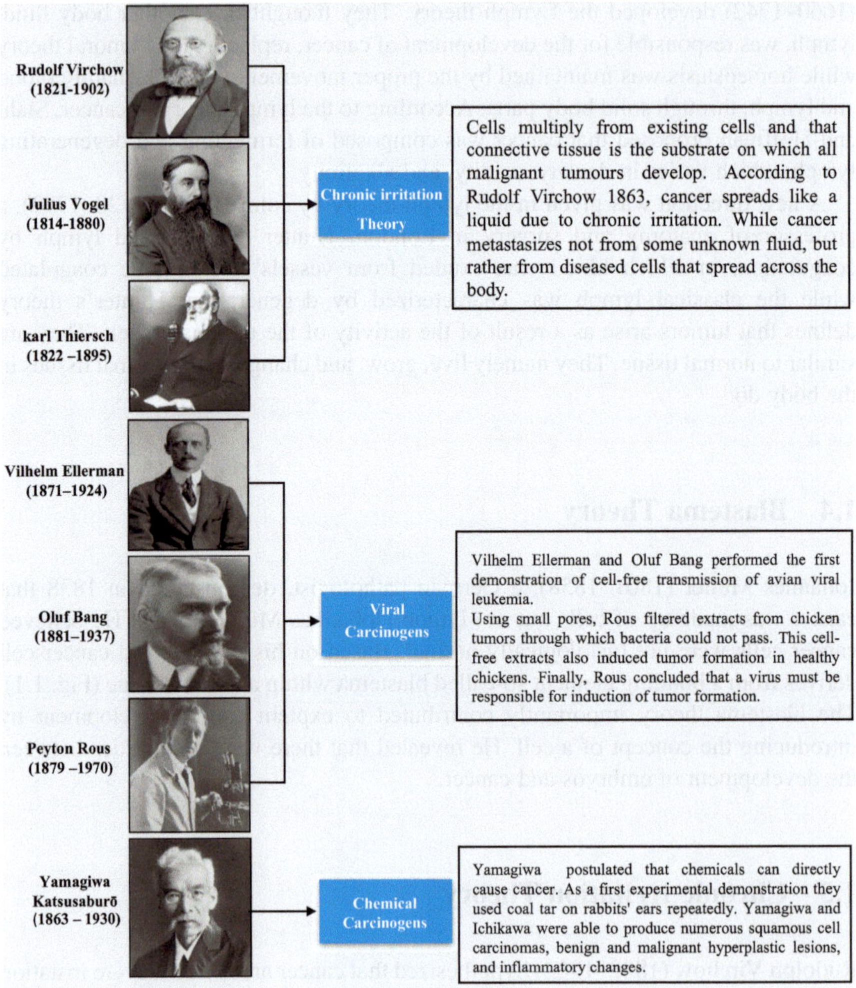

Fig. 1.3 Historical view on the theories of causes of cancer

His views on dyscrasia were similar to those of Hippocrates, who taught that dyscrasia explains all diseases according to a person's constitution and disposition. However, Karl Thiersch, a German surgeon, discovered in the 1860s that cancer metastasis was not from some unknown fluid, but rather from diseased cells that spread across the body.

Johannes Andreas Grib Fibiger (1867–1928), the Danish pathologist, experimentally showed worm larvae, named *Gongylonema neoplasticum*, infected from ingested cockroaches caused stomach cancer in rats (Fibiger 1913). However, worm larvae were later found to induce only benign tumor and vitamin A deficiency as the primary cause of cancer.

In 1915, two Japanese pathologists at the University of Tokyo, Katsusaburo Yamagiwa (1863–1930) and Koichi Ichikawa (1888–1948), demonstrated the first experimental induction of cancer (Yamagiwa and Ichikawa 1915). It was also considered as the first experiment to demonstrate coal tar as a chemical carcinogen. However, coal tar is nowadays believed not to contain a carcinogen but irritants.

Painting coal tar on rabbits' ears repeatedly and inducing chronic inflammation, Yamagiwa and Ichikawa were able to produce numerous squamous cell carcinomas, benign and malignant hyperplastic lesions. They also described the transformation of benign cells into malignant ones and the regression of benign tumors as part of the carcinogenic process for the first time (Yamagiwa and Ichikawa 1915).

1.6 Viral Carcinogenesis

In 1908, Vilhelm Ellermann (1871–1924) and Olaf Bang (1877–1953) demonstrated that cell-free filtrate of chicken leukemic cell extracts could transmit the disease to healthy chickens (Ellermann and Bang 1908). However, the leukemia cell extracts were not classified as tumorigenic until when the avian leukemia virus was identified. It took another 40 years before leukemia was accepted as a cancer derived from bone marrow (Epstein 2001). Therefore, Ellermann and Bang's important discovery was largely kept unnoticed for a first half of the twentieth century.

An American scientist Francis Peyton Rous (1879–1970) discovered the tumor-causing retro virus which is now known as Rous sarcoma virus. Rous was studying the transmission of tumors in chickens. He found that a tumor was induced in a healthy chicken, which had small pieces of a tumor from a cancer-prone chicken (Rous 1910). Then he made extracts from the chicken tumors and filtered them through small pores that blocked the passage of bacteria. Interestingly, these cell-free extracts also induced tumors in healthy chickens (Rous 1910). Because filtration was not effective to exclude viruses, Rous concluded that a virus should be responsible for the induction of tumor.

In 1975, Michael Bishop and Harold Varmus identified an oncogene responsible for causing cancer in the genome of Rous sarcoma virus and found it in fact copied from the host to the virus. The oncogene in the host cell was called *c*-src and classified as proto-oncogene while the oncogene derived from Rous sarcoma virus was called *v*-src defined as viral oncogene. Later, they showed that the c-src gene was normally involved in the positive regulation of the host cell growth and differentiation. Once infected, retro virus genome is reverse transcribed into DNA and integrated into host genome by the long terminal repeat allowing the overexpression of v-src in the host cell resulting in unlimited host cell growth leading to cancer.

In 1965, Michael Anthony Epstein (1921–present), Yvonne Barr (1932–2016), and Bert Geoffrey Achong (1928–1996) visualized herpes virus-like particles in a cell line established from Burkitt's lymphoma (BL) by electron microscopy (Epstein et al. 1965). This DNA virus was found to be biologically and antigenically distinct from other known human herpes viruses (Henle and Henle 1966) and was named the Epstein–Barr Virus (EBV). The viral genome encode EBNA-2, EBNA-3C, and LMP-1 proteins are essential for the transformation of host cells.

In 1967 and 1968, Baruch Samuel Blumberg (1925–2011), Kazuo Okochi, Shinya Murakami, and Alfred M. Prince (1928–2011) described that blood from hepatitis patients contained the Australia antigen (Okochi and Murakami 1968; Prince 1968), which was the surface antigen of a hepadnavirus called hepatitis B virus (HBV) (Ganem and Schneider 2001). In 1975, Blumberg and colleagues linked chronic HBV infection to hepatocellular carcinoma (HCC) (Blumberg et al. 1975), which was one of the most frequent cancers all over the world. Importantly, in 1976 the first effective HBV vaccine was developed by large-scale purification of HBV surface antigen from the serum of HBV carriers (Buynak et al. 1976) followed by a second-generation vaccine utilizing recombinant HBV surface antigen, which has been produced since the 1980s and is still in use. Since HBV vaccine protects people from not only acute and chronic hepatitis but the development of HCC (Hilleman 2003; Chang et al. 1997), HBV is one of the keys to approach in liver cancer research.

The discovery of several oncogenic viruses has been a significant advancement in cancer research. Pathological and genomic approaches also contributed to understand RNA and DNA viruses (e.g., RSV, SV40, HIV/lentivirus, and adenoviruses) that episomally replicate or integrate into host genomes resulting in the induction of tumors in animals known as transformation of cells in vitro. The relatively small genome sizes of these viruses have enabled the development of molecular paradigms beyond virology, leading to current molecular medicine (Butel 2000).

1.7 Mutation Theory

Theodor Boveri (1862–1915) (Fig. 1.4) wrote in his book that cancer was a cell-based disease, and that the problem of tumors was a cell problem. Also, he described that cancers occurred due to an abnormal chromosomal rearrangement that eliminated a portion of chromosomal material whose function was to inhibit cell proliferation. Finally, Boveri theorized that the cause of cancer was not abnormal mitosis but a certain abnormal chromatin-complex, no matter how it arises (Boveri 1914). Boveri was not confident with his idea on carcinogenesis, because few colleagues stood for him when he first introduced his theory in his book. His wife, Marcella first translated his book into English in 1929 (Oppenheimer 1967).

"Somatic mutation" was first coined by Ernest E. Tyzzer (1875–1965) in 1916 (Tyzzer 1916). The view of somatic mutations in cancer was further expanded in 1919 when Whitman proposed that a cancer cell was a "mutated" cell and explained what Boveri had implied (Whitman 1919; Wunderlich 2007). Then in 1928, Karl-Heinrich Bauer (1890–1978) suggested that cancer could be caused by mutations (Bauer 1928). During a century up to now, the concept of cancer has been developed and modified based on Boveri's original somatic mutation theory (SMT) (Fig. 1.3). After the DNA structure was identified as a double helix and genetic information was

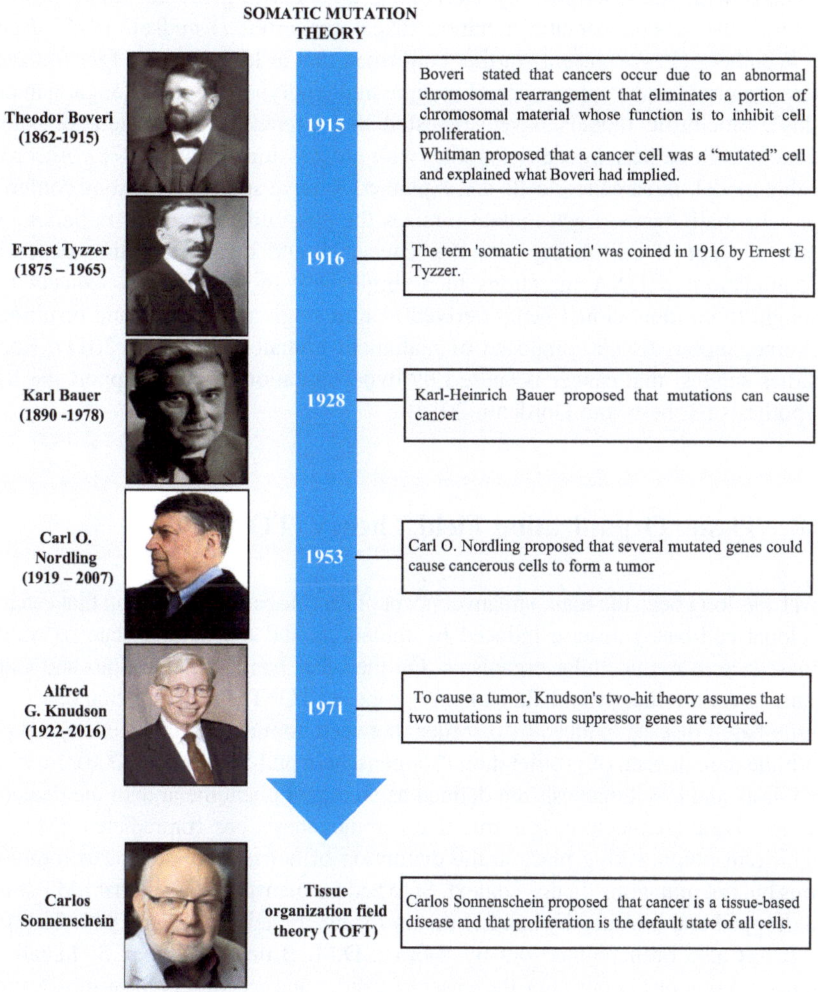

Fig. 1.4 Historical view of cancer causes through SMT to TOFT

discovered in the middle of the twentieth century (Cobb 2013), Carl O. Nordling (1919–2007) simultaneously suggested that several mutated genes could cause cancerous cells to form a tumor (Nordling 1953).

After this hypothesis, the number of mutational changes required to cause cancer was extensively investigated. As a result, Ashley suggested that approximately three to seven mutations might be required for the development of cancer (Ashley 1969).

One-hit carcinogenesis is a model of dose-response that assumes that one genetic change transforms a normal cell into a cancerous one, and that any dose of a carcinogen is harmful. The risk of harm can only be completely eliminated at zero dose. Alfred G. Knudson, Jr. (1922–2016) modified this mutation theory eventually replaced with the "two-hit" by explaining an inheriting mutant allele needs to undergo the second somatic mutation to get cancerous (Knudson 1971). Robert A. Weinberg (1942–present) further confirmed that at least three or four mutations were required for the appearance of malignant phenotypes in vitro (Rangarajan et al. 2004). Stochastic models have suggested that events of serial mutation should generate cellular heterogeneity together with progression (Nowell 1976). According to this model, most cancer cells are explained to have several mutations conferring the cells malignancy. Each mutation raises the possibility of cancer initiation. The main concept of this theory was that cancer should result from time-dependent accumulation of DNA mutations in a single cell. In this context, cancers were thought to be monoclonal being derived from a single mutant cell and resulting in a homogeneous tissue composed of malignant mutant cells (Vaux 2011). Recent studies suggest that cancer is caused by hyper-mutations, which support the SMT hypothesis (Roberts and Gordenin 2014).

1.8 Tissue Organization Field Theory (TOFT)

SMT has long been the main-stream concept of carcinogenesis assuming that cancer is a clonal cell-based disease induced by mutations and the default state of cells as quiescence in multicellular organisms. On the other hand, Ana M. Soto and Carlos Sonnenschein critically and alternatively proposed TOFT conceptualizing cancer as a tissue-based disease induced by disorders between normal stromal/parenchymal cells with the default state of proliferation (Sonnenschein and Soto 1999, 2000). In TOFT, neoplasia and carcinogenesis are defined as emergent phenomena with the defects of normal tissue architecture. It is true that carcinogenesis has contradicted SMT with significant points taking place at the disruption of normal intercellular communications but not mutations. In this context, Soto and Sonnenschein assert that SMT should be dropped and replaced by another theory (Sonnenschein and Soto 1999, 2000).

It has also been pointed out by Bjorn L.D.M. Brücher and Ijaz S. Jamall that somatic mutations occur after the onset of cancer and are observed in most cancers after recognizing the early cues of carcinogenesis (Brücher and Jamall 2016). According to them, the origin of the vast majority of cancers as clinical data shows little support for the SMT when compared to patient outcomes. Since the SMT is not always applicable to all cancers, and even to chemical carcinogenesis, in the absence of hard evidence of causality, they propose a new paradigm, which is scientifically applicable for the majority of nonheritable cancers, consisting of a six-step sequence through chronic inflammation for the origin of cancer.

1.9 Stem Cells

The term "stem cell" was coined in 1868 by a German biologist, Ernst Haeckel (1834–1919) (Fig. 1.5), by describing the fertilized egg or an undifferentiated embryo (Haeckel 1868; Ramalho-Santos and Willenbring 2007; Cooper 2009).

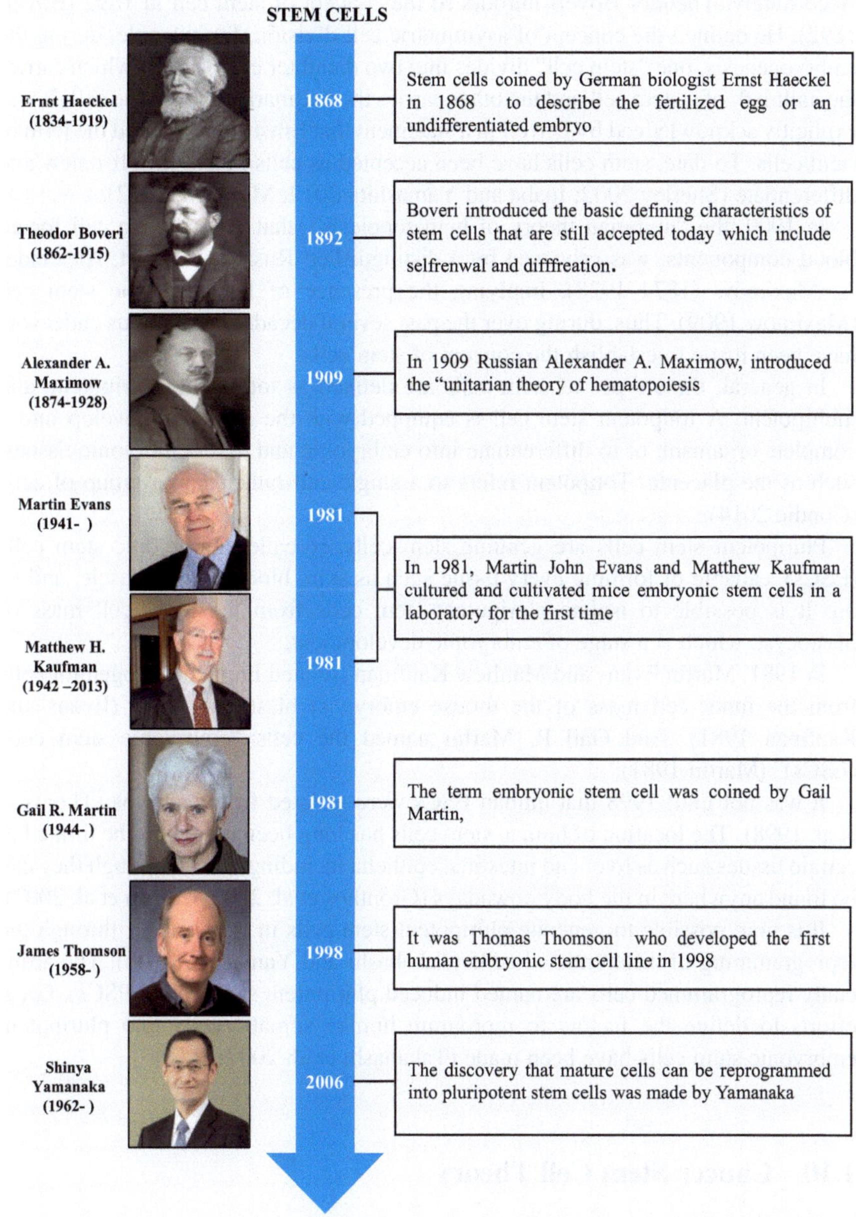

Fig. 1.5 The historical view on the progress of stem cell research: concept, identification, and preparation

Accordingly, Theodor Boveri introduced the concept of stem cell in 1892 (Boveri 1892). He defined the concept of asymmetric cell division, for example, during the embryogenesis, one "stem cell" divides into two daughter cells, one of which carries the hallmark of a stem cell and the other carries the hallmark of a somatic cell. It was explicitly acknowledged by Boveri in a statement that Ernst Haeckel used the term of stem cells. To date, stem cells have been accepted as cells potent to self-renew and differentiate (Sherley 2002; Inaba and Yamashita 2012; Monti M. 2012).

In 1909, the "unitarian theory of hematopoiesis", that is, one stem cell for all blood components, was proposed by a distinguished Russian scientist, Alexander A. Maximow (1874–1928), implying the presence of hematopoietic stem cell (Maximow 1909). Thus, during over the past several decades, continuous endeavors have been made to establish the concept of stem cells.

In general, three types of stem cells are defined as totipotent, pluripotent, and multipotent. A totipotent stem cell is equipped with the ability to develop into a complete organism, or to differentiate into embryonic and extra-embryonic tissues, such as the placenta. Totipotent refers to a single cell rather than a group of cells (Condic 2014).

Pluripotent stem cells are genuine stem cells, so-called embryonic stem cells (ESCs), capable of forming every tissue such as skin, blood, bone, muscle, and so on. It is possible to isolate pluripotent stem cells from the inner cell mass of blastocyst, which is a stage of embryonic development.

In 1981, Martin Evans and Matthew Kaufman isolated immature progenitor cells from the inner cell mass of the mouse embryo at blastocyst stage (Evans and Kaufman 1981). And Gail R. Martin named the cells "embryonic stem cells (ESCs)" (Martin 1981).

It was not until 1998 that human ESCs were isolated from embryos (Thomson et al. 1998). The location of human stem cells has long been thought to be limited to certain tissues such as liver and intestinal epithelia including blood although they can be found anywhere in the body nowadays (Gronthos et al. 2000; Cregan et al. 2007).

It is now possible to generate pluripotent stem cells in a laboratory through the reprogramming of differentiated cells (Takahashi and Yamanaka 2006). The artificially reprogrammed cells are named induced pluripotent stem cells (iPSCs). Great efforts to define the factors to reprogram human somatic cells into pluripotent embryonic stem cells have been made (Takahashi et al. 2007).

1.10 Cancer Stem Cell Theory

In the mid-nineteenth century, tumor growth was hypothesized to be driven by a small number of cells, so-called cancer stem cells (CSCs). The research in stem cells may now drop the mutation theory from the main context of cancer development.

While Müller was the first to describe cancer as the abnormal continuation of embryonic cell development in 1838 (Fig. 1.6) (Müller 1838), the first mention of

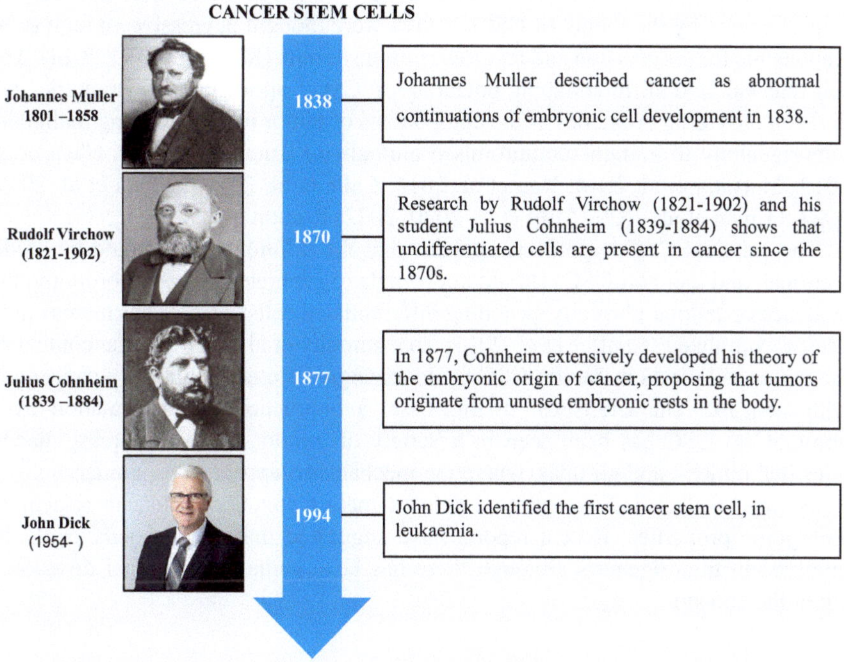

Fig. 1.6 The historical view on the progress of cancer stem cell research

undifferentiated cells in cancer was not until the 1870s during the research by Virchow and his student Julius Cohnheim (1839–1884).

In 1877, Cohnheim proposed a theory of cancer development as the embryonic origin (Cohnheim 1877). In this theory, tumors were hypothesized as a result of "embryonic rests" occurring within the body (Cooper 2009). Beyond the morphological similarities, Cohnheim tried to postulate the existence of embryonic cell remnants, which uncontrollably proliferated receiving the necessary blood supply due to their embryonic character, in the body as the cause of all tumors. Accordingly, he thought tumors would develop due to the "errors" occurring in the embryonic development. In addition, Cohnheim noted that normal cell growth might be promoted with these embryonic cells during puberty or pregnancy (Capp 2019).

The first evidence for the CSCs was provided by John Dick and his co-workers (Lapidot et al. 1994; Bonnet and Dick 1997). Using FACS, they isolated the CSCs residing in human acute myeloid leukemia (AML) cells that initiated leukemia when they were transplanted mice.

Since CSCs and normal stem cells were found similar in terms of self-renewal potential and pluripotency (Al-Hajj and Clarke 2004; Al-Hajj et al. 2003), a new concept of cancer development has been hypothesized by abnormal stem cell biology based on epigenetics rather than mutations (Wicha et al. 2006; Afify and Seno 2019; Afify et al. 2022).

Only certain subpopulations of cancer cells were suggested to drive the progression of cancer by the model of CSCs, which were thought aggressive enough to be responsible for progression and recurrence of the tumors (Kreso and Dick 2014). The self-renewal and differentiation potential of CSCs allow their asymmetric cell division providing progenies as the components of tumor mass exhibiting malignant tumorigenicity in immune-compromised animals (O'Brien et al. 2007, Chen et al. 2012, Matsuda et al. 2014, Yan et al. 2014, Calle et al. 2016, Hassan et al. 2019, 2022, Osman et al. 2020, Afify et al. 2020, 2022, Kumon et al. 2021).

The classical CSC hypothesis explains that the differentiation should be unidirectional, and that non-CSCs should not be able to generate CSCs. Accordingly, the evidence of cellular plasticity switching differentiated cells to CSCs has increasingly been accumulated (Chaffer et al. 2011; Thankamony et al. 2020). In the context of the tumor cell plasticity, a non-CSC could undergo differentiation into a cancer cell with stem-like characteristics. In this case, a bidirectional transformation from non-CSC to CSC has been seen in a variety of tumors, including breast cancer, intestinal tumors, and gliomas, where the mechanisms appear to be similar.

When a cell is dedifferentiated, it loses mature functions and can re-acquire embryonic properties. Recent reports have suggested that the plasticity could be involved in tumorigenesis although there has been some controversial discussion about the concept.

1.11 Origin of the Cancer Stem Cell

Concerning the origin of CSCs, multiple hypotheses have been proposed. They are feasible but still puzzling.

1.11.1 Cell Fusion

In the process of cell fusion, many uninucleate cells (cells with one nucleus) combine to form multinucleate cells, known as syncytia. The fusion of cells takes place during the differentiation, embryogenesis, morphogenesis, as well as viral infection (Shemer and Podbilewicz 2003; Ogle et al. 2005). Since it is common for the nuclei of the fusion partners to remain separate, such as in placenta and muscles, two somatic cells will basically produce a tetraploid hybrid.

However, cancer initiation will be possible when two or more nuclei are accidentally fused under abnormal conditions (Fujiwara et al. 2005). There is increasing evidence that abnormal fusion between cells should contribute to the progression of cancer. Cellular fusion between stem cells and preexisting differentiated cells may occur resulting in transdifferentiation, which challenges the plasticity of stem cells. In this context, the initiation of cancer could be correlated with the increasing frequency of cell-cell fusion. The fusion of tumor cells with normal somatic cells often produces hybrid cells that are more malignant than the parents (Duelli and Lazebnik 2003; Pawelek 2000). However, the cell-cell fusion still sounds not enough to explain the appearance of CSCs.

Aractingi et al. tried to link stem cells to the appearance of CSC by cell-cell fusion and reported that stem cells could migrate to the skin, differentiate or fuse into keratinocytes, and undergo transformation (Aractingi et al. 2005). Simultaneously, Bjerkvig et al. hypothesized that the fusion of an adult stem cell, even if it is somatic or hematopoietic, with a differentiated cell could lead to form a CSC (Bjerkvig et al. 2005). The fusion of bone-marrow-derived cells with differentiated adult tissue cells dramatically increased under chronic inflammation, which is a major risk factor for tumorigenesis (Johansson et al. 2008; Nygren et al. 2008).

A unique set of cell survival programs could also exist in the stem cells, which may be responsible for driving tumor initiation. And hence, there could be a possibility that these fused cells retain stem-cell features with large chromosomal aberrations. Cancer development may be further explained by the tumor initiation of the fusion between stem cells and somatic cells along with a series of mutational events observed as chromosomal derangements.

1.11.2 Mutation

Cancer development including tumor initiation has majorly been believed to be responsible for mutations. According to the mutation theory, mutations in stem cells, progenitor cells, mature cells, or cancer cells should induce CSCs. The self-renewal and differentiation potentials in CSC are hypothesized to be controlled by genetic programs.

In this context, the transformation of stem cell into CSCs along with aging could result from irreversible changes such as accumulated nuclear and mitochondrial DNA damages, and shortening telomeres (Pollina and Brunet 2011). Mutations and/or loss of tumor suppressor genes like p21 or Tp53 are the representatives (Rodriguez et al. 2009). The rate of DNA mutation increases in expanding stem cells at high oxygen concentration during replication should lead to CSC induction (Bétous et al. 2016).

1.11.3 Epigenetics

1.11.3.1 Non-stem Cancer Cells

Yanger et al. suggested that CSCs appear from non-stem cancer cells by the upregulation of pluripotency-related pathways such as JAGGED2-NOTCH, WNT/β-CATENIN, and Hedgehog (Yanger et al. 2013). Dedifferentiation of mature cells into CSC-like phenotypes has also been demonstrated by inactivating the

Hippo pathway, a regulator of morphogenesis (Yimlamai et al. 2014). The stemness has also been reported to be imparted in non-stem cancer cells by sufficient amount of IL6 (Iliopoulos et al. 2011). Widschwendter et al. discovered that polycomb group proteins were permanently suppressing the genes necessary for differentiation by hypermethylation in cancer cells allowing continuous self-renewal (Widschwendter et al. 2007). Endothelial nitric oxide was reported to activate NOTCH signaling promoting stem cell-like character in the tumor (Charles et al. 2010). Hypoxia in cancer allowed the accumulation of transcription factors HIF1α and HIF2α to induce NOTCH ligands to induce stemness as well as EMT-related genes such as SNAIL, SLUG, TWIST, and N-cadherin (Jögi et al. 2002; Xing et al. 2011). Further, endothelial cells expressing Sonic Hedgehog also promoted Gli1 positive glioma cells to invoke stemness through Hedgehog pathway (Yan et al. 2014).

1.11.3.2 Stem Cells

For more than a decade our group has demonstrated the generation of CSCs from stem cells under the condition of chronic inflammation without any genetic manipulation (Chen et al. 2012; Yan et al. 2014; Calle et al. 2016; Afify et al. 2020; Minematsu et al. 2022). Chronic exposure to prostaglandin E2 was found effective to induce CSCs, in which PI3K/AKT signaling pathway was enhanced. This activation was implied by the epigenetic overexpression of the moieties of PI3K due to the hypomethylation of CpG islands (Oo et al. 2018; Afify et al. 2021).

This generation of CSCs was also demonstrated with tissue specificity such as breast cancer (Abu Quora et al. 2021), pancreatic cancer in vitro (Calle et al. 2016) or in vivo (Hassan et al. 2022), and liver (Afify et al. 2020).

The generated CSCs also showed the development of tumor microenvironment by the differentiation potential to endothelial cells (Matsuda et al. 2014, Prieto-Vila et al. 2016), cancer-associated fibroblast (Nair et al. 2017), tumor-associated macrophage (Osman et al. 2020), and hematopoietic stem cells (Hassan et al. 2019), including erythrocytes (Kumon et al. 2021). These findings could explain the heterogeneity of a tumor tissue and cellular hierarchy.

Considering the process of CSC generation through our studies, we hypothesize the cancer-inducing niche as the microenvironment of tumor initiation, which means the change of normal cells into CSCs. Cancer-inducing niche explains chronic inflammation as the long-term exposure of pro-inflammatory mediators and growth factors to the normal stem and progenitor cells, which acquire the phenotype of CSC as the result (Afify et al. 2022, 2022).

References

Abu Quora HA, Zahra MH, El-Ghlban S, Nair N, Afify SM, Hassan G, Nawara HM, Sheta M, Monzur S, Fu X, Osman A, Seno A, Seno M. Microenvironment of mammary fat pads affected the characteristics of the tumors derived from the induced cancer stem cells. Am J Cancer Res. 2021 Jul 15;11(7):3475–95.

Afify SM, Hassan G, Seno A, Seno M. Cancer-inducing niche: the force of chronic inflammation. Br J Cancer. 2022 Jul;127(2):193–201.
Afify SM, Hassan G, Yan T, Seno A, Seno M. Cancer stem cell initiation by tumor-derived extracellular vesicles. Methods Mol Biol. 2022;2549:399–407.
Afify SM, Oo AKK, Hassan G, Seno A, Seno M. How can we turn the PI3K/AKT/mTOR pathway down? Insights into inhibition and treatment of cancer. Expert Rev Anticancer Ther. 2021 Jun;21(6):605–19.
Afify SM, Sanchez Calle A, Hassan G, Kumon K, Nawara HM, Zahra MH, Mansour HM, Khayrani AC, Alam MJ, Du J, Seno A, Iwasaki Y, Seno M. A novel model of liver cancer stem cells developed from induced pluripotent stem cells. Br J Cancer. 2020 Apr;122(9):1378–90.
Afify SM, Seno M. Conversion of stem cells to cancer stem cells: undercurrent of cancer initiation. Cancers (Basel). 2019 Mar 11;11(3):345.
Al-Hajj M, Clarke MF. Self-renewal and solid tumor stem cells. Oncogene. 2004;23:7274–82.
Al-Hajj M, Wicha MS, Benito-Hernandez A, Morrison SJ, Clarke MF. Prospective identification of tumorigenic breast cancer cells. Proc Natl Acad Sci U S A. 2003;100:3983–8.
Aractingi S, et al. Skin carcinoma arising from donor cells in a kidney transplant recipient. Cancer Res. 2005;65:1755–60.
Ashley DJB. The two "hit" and multiple "hit" theories of carcinogenesis. Br J Cancer. 1969;23: 313–28.
Bauer KH. Mutationstheorie der Geschwulst-Entstehung. Berlin: Julius Springer Verlag; 1928.
Bétous R, Renoud ML, Hoede C, Gonzalez I, Jones N, et al. (2016) Human adipose-derived stem cells expanded under ambient oxygen concentration accumulate oxidative DNA lesions and experience Procarcinogenic DNA replication stress. Stem Cells Transl Med. 6(1):68–76.
Bjerkvig R, Tysnes BB, Aboody KS, Najbauer J, Terzis AJ. Opinion: the origin of the cancer stem cell: current controversies and new insights. Nat Rev Cancer. 2005 Nov;5(11):899–904.
Blumberg BS, Larouze B, London WT, Werner B, Hesser JE, Millman I, Saimot G, Payet M. The relation of infection with the hepatitis B agent to primary hepatic carcinoma. Am J Pathol. 1975;81:669–82.
Bonnet D, Dick JE. Human acute myeloid leukemia is organized as a hierarchy that originates from a primitive hematopoietic cell. Nat Med. 1997;3:730–7.
Boveri T. Ueber die Entstehung des Gegensatzes zwischen den Geschlechtszellen und den somatischen Zellen bei Ascaris megalocephala, nebst Bemerkungen zur Entwicklungsgeschichte der Nematoden. Sitzungsberichte der Gesellschaft für Morphologie und Physiologie in München. 1892;8:114–25.
Boveri T. Zur Frage der Entstehung maligner Tumoren. Germany: Verlag von Gustav Fischer; Jena; 1914. p. 29–32.
Breasted JH. The Edwin Smith surgical papyrus, Volume 1: hieroglyphic transliteration, translation, and commentary. 1930. (Reissued 1991 with a foreword by T. A. Holland).
Brücher BL, Jamall IS. Somatic mutation theory–why it's wrong for Most cancers. Cell Physiol Biochem. 2016;38(5):1663–80.
Butel JS. Viral carcinogenesis: revelation of molecular mechanisms and etiology of human disease. Carcinogenesis. 2000;21:405–26.
Buynak EB, Roehm RR, Tytell AA, Bertland AU, Lampson GP, Hilleman MR. Vaccine against human hepatitis B. JAMA. 1976;235:2832–4.
Calle AS, Nair N, Oo AK, Prieto-Vila M, Koga M, Khayrani AC, Hussein M, Hurley L, Vaidyanath A, Seno A, Iwasaki Y, Calle M, Kasai T, Seno M. A new PDAC mouse model originated from iPSCs-converted pancreatic cancer stem cells (CSCcm). Am J Cancer Res. 2016 Dec 1;6(12):2799–815.
Capp JP. Cancer stem cells: from historical roots to a new perspective. J Oncol. 2019 Jun;11(2019): 5189232.
Castiglioni A. Histoire de la Medecine. Paris: Payot; 1931.

Chaffer CL, Brueckmann SC, Kaestli AJ, Wiggins PA, Rodrigues LO, Brooks M, Reinhardt F, Su Y, Polyak K, et al. Normal and neoplastic nonstem cells can spontaneously convert to a stem-like state. Proc Natl Acad Sci U S A. 2011;108:7950–5.

Chang MH, Chen CJ, Lai MS, Hsu HM, Wu TC, Kong MS, Liang DC, Shau WY, Chen DS. Universal hepatitis B vaccination in Taiwan and the incidence of hepatocellular carcinoma in children. Taiwan childhood Hepatoma study group. N Engl J Med. 1997;336:1855–9.

Charles N, Ozawa T, Squatrito M, Bleau AM, Brennan CW, et al. Perivascular nitric oxide activates notch signaling and promotes stem-like character in PDGF-induced glioma cells. Cell Stem Cell. 2010;6:141–52.

Chen L, Kasai T, Li Y, Sugii Y, Jin G, Okada M, Vaidyanath A, Mizutani A, Satoh A, Kudoh T, Hendrix MJ, Salomon DS, Fu L, Seno M. A model of cancer stem cells derived from mouse induced pluripotent stem cells. PLoS One. 2012;7(4):e33544.

Cobb M. When genes become "information". Cell. 2013;153:503–6.

Cohnheim J. Vorlesungen über allgemeine Pathologie (Erster Band). 2nd ed. Berlin: Hirschwald; 1877(1882).

Condic ML. Totipotency: what it is and what it is not. Stem Cells Dev. 2014;23(8):796–812. https://doi.org/10.1089/scd.2013.0364. Epub 2014 Feb 12

Cooper M. Regenerative pathologies: stem cells, teratomas and theories of cancer. Medicine Studies. 2009;1(1):55–66.

Cregan MD, Fan Y, Appelbee AJ, Brown ML, Klopcic B, Koppen JA, Mitoulas LR, Piper KME, Choolani MA, Chong YS, et al. Identification of nestin-positive putative mammary stem cells in human breast milk. Cell Tissue Res. 2007;329:129–36.

Diamandopoulos GT. Cancer: an historical perspective. Anticancer Res. 1996;16:1595–602.

Duelli D, Lazebnik Y. Cell fusion: a hidden enemy? Cancer Cell. 2003;3:445–8.

Ebbell B. The Papyrus Ebers. The greatest Egyptian medical document. 135. London: Humphrey Milford; 1937. p. 15s.

Ellermann V, Bang O. Experimentelle leukamie bei huhnern. Zentralbl Bakteriol Parasitenkd Infektionskr Hyg. 1908;46:595–7.

Epstein MA. Historical background. Philos Trans R Soc Lond Ser B Biol Sci. 2001;356:413–20.

Epstein MA, Henle G, Achong BG, Barr YM. Morphological and biological studies on a virus in cultured lymphoblasts from Burkitt's lymphoma. J Exp Med. 1965;121:761–70. https://doi.org/10.1084/jem.121.5.761.

Evans M, Kaufman M. Establishment in culture of pluripotent cells from mouse embryos. Nature. 1981;292:154–6.

Fibiger JAG. Untersuchungen über eine Nematode spiroptera sp. n. und deren Fähigkeit papillomatöse und carcinomatöse Geschwulstbildungen im Magen der Ratte hervorzurufen. Z Krebsforschung. 1913;13:217–80.

Fujiwara T, Bandi M, Nitta M, Ivanova EV, Bronson RT, Pellman D. Cytokinesis failure generating tetraploids promotes tumorigenesis in p53-null cells. Nature. 2005;437(7061):1043–7.

Ganem D, Schneider R. Hepadnaviridae: the viruses and their replication. Fields virology. In: Knipe DM, Howley PM, editors. , vol. 2. Philadelphia: Lipppincott Williams & Wilkins; 2001. p. 2923–70.

Gronthos S, Mankani M, Brahim J, Robey PG, Shi S. Postnatal human dental pulp stem cells (DPSCs) in vitro and in vivo. Proc Natl Acad Sci U S A. 2000;97:13625–30.

Haeckel E. Natürliche Schöpfungsgeschichte. Georg Reimer: Berlin, Germany; 1868.

Hajdu SI. Greco-Roman thought about cancer. Cancer. 2004;100:2048–51.

Hajdu SI. A note from history: landmarks in history of cancer, part 1. Cancer. 2011 Mar 1;117(5): 1097–102.

Hassan G, Afify SM, Nair N, Kumon K, Osman A, Du J, Mansour H, Abu Quora HA, Nawara HM, Satoh A, Zahra MH, Okada N, Seno A, Seno M. Hematopoietic cells derived from cancer stem cells generated from mouse induced pluripotent stem cells. Cancers (Basel). 2019 Dec 29;12(1): 82.

Hassan G, Ohara T, Afify SM, Kumon K, Zahra MH, Fu X, Al Kadi M, Seno A, Salomon DS, Seno M. Different pancreatic cancer microenvironments convert iPSCs into cancer stem cells exhibiting distinct plasticity with altered gene expression of metabolic pathways. J Exp Clin Cancer Res. 2022 Jan 21;41(1):29.

Henle G, Henle W. Immunofluorescence in cells derived from Burkitt's lymphoma. J Bacteriol. 1966;91:1248–56.

Hilleman MR. Critical overview and outlook: pathogenesis, prevention, and treatment of hepatitis and hepatocarcinoma caused by hepatitis B virus. Vaccine. 2003;21:4626–49.

Iliopoulos D, Hirsch HA, Wang G, Struhl K. Inducible formation of breast cancer stem cells and their dynamic equilibrium with non-stem cancer cells via IL6 secretion. Proc Natl Acad Sci U S A. 2011;108:1397–402.

Inaba M, Yamashita YM. Asymmetric stem cell division: precision for robustness. Cell Stem Cell. 2012;11:461–9.

Javier RT, Butel JS. The history of tumor virology. Cancer Res. 2008 Oct 1;68(19):7693–706.

Jögi A, Øra I, Nilsson H, Lindeheim A, Makino Y, et al. Hypoxia alters gene expression in human neuroblastoma cells toward an immature and neural crest-like phenotype. Proc Natl Acad Sci U S A. 2002;99:7021–6.

Johansson CB, Youssef S, Koleckar K, et al. Extensive fusion of hematopoietic cells with Purkinje neurons in response to chronic inflammation. Nat Cell Biol. 2008;10(5):575–83.

Kardinal CG, Yarbro JW. A conceptual history of cancer. Semin Oncol. 1979;6:396–408.

Knudson AG. Mutation and cancer: statistical study of retinoblastoma. Proc Natl Acad Sci U S A. 1971;68:820–3.

Kreso A, Dick JE. Evolution of the cancer stem cell model. Cell Stem Cell. 2014;14:275–91.

Kumon K, Afify SM, Hassan G, Ueno S, Monzur S, Nawara HM, Quora HAA, Sheta M, Xu Y, Fu X, Zahra MH, Seno A, Seno M. Differentiation of cancer stem cells into erythroblasts in the presence of CoCl2. Sci Rep. 2021 Dec 14;11(1):23977.

Lapidot T, Sirard C, Vormoor J, et al. A cell initiating human acute myeloid leukaemia after transplantation into SCID mice. Nature. 1994;367:645–8.

Major RH. A history of medicine. Springfield: Ch. C. Thomas; 1954.

Martin GR. Isolation of a pluripotent cell line from early mouse embryos cultured in medium conditioned by teratocarcinoma stem cells. Proc Natl Acad Sci U S A. 1981;78:7634–8.

Matsuda S, Yan T, Mizutani A, Sota T, Hiramoto Y, Prieto-Vila M, Chen L, Satoh A, Kudoh T, Kasai T, Murakami H, Fu L, Salomon DS, Seno M. Cancer stem cells maintain a hierarchy of differentiation by creating their niche. Int J Cancer. 2014 Jul 1;135(1):27–36.

Maximow AA. Der Lymphozyt als gemeinsame Stammzelle verschiedenen Blutelemente in der embryonalen Entwicklung und im postfetalen Leben der Säugetiere. Folia. Haematologica. 1909;8:125–34.

Minematsu H, Afify SM, Sugihara Y, Hassan G, Zahra MH, Seno A, Adachi M, Seno M. Cancer stem cells induced by chronic stimulation with prostaglandin E2 exhibited constitutively activated PI3K axis. Sci Rep. 2022 Sep 17;12(1):15628.

Monti M, Perotti C, Del Fante C, Cervio M, Redi CA, Fondazione IRCCS Policlinico San Matteo, Pavia (Italia). Stem cells: sources and therapies. Biol Res. 2012;45:207–14.

Müller J. Über den feinern Bau und die Formen der krankhaften Geschwülste. Berlin: G. Reimer; 1838.

Nair N, Calle AS, Zahra MH, et al. A cancer stem cell model as the point of origin of cancer-associated fibroblasts in tumor microenvironment. Sci Rep. 2017;7(1):6838. Published 2017 Jul 28. https://doi.org/10.1038/s41598-017-07144-5.

Nordling CO. A new theory on the cancer-inducing mechanism. Br J Cancer. 1953;7:68–72.

Nowell PC. The clonal evolution of tumor cell populations. Science. 1976;194:23–8.

Nygren JM, Liuba K, Breitbach M, et al. Myeloid and lymphoid contribution to non-hematopoietic lineages through irradiation-induced heterotypic cell fusion. Nat Cell Biol. 2008;10(5):584–92.

O'Brien CA, Pollett A, Gallinger S, Dick JE. A human colon cancer cell capable of initiating tumour growth in immunodeficient mice. Nature. 2007;445:106–10.

Ogle BM, Cascalho M, Platt JL. Biological implications of cell fusion. Nature Rev Mol Cell Biol. 2005;6:567–75.
Okochi K, Murakami S. Observations on Australia antigen in Japanese. Vox Sang. 1968;15:374–85.
Oo AKK, Calle AS, Nair N, Mahmud H, Vaidyanath A, Yamauchi J, Khayrani AC, Du J, Alam MJ, Seno A, Mizutani A, Murakami H, Iwasaki Y, Chen L, Kasai T, Seno M. Up-regulation of PI 3-kinases and the activation of PI3K-Akt signaling pathway in cancer stem-like cells through DNA Hypomethylation mediated by the cancer microenvironment. Transl Oncol. 2018 Jun;11 (3):653–63.
Oppenheimer J. Essays in the history of embryology. Cambridge, MA: MIT Press; 1967. p. 80.
Osman A, Oze M, Afify SM, Hassan G, El-Ghlban S, Nawara HM, Fu X, Zahra MH, Seno A, Winer I, Salomon DS, Seno M. Tumor-associated macrophages derived from cancer stem cells. Acta Histochem. 2020 Dec;122(8):151628.
Pawelek JM. Tumour cell hybridization and metastasis revisited. Melanoma Res. 2000;10:507–14.
Pollina EA, Brunet A. Epigenetic regulation of aging stem cells. Oncogene. 2011;30:3105–26.
Prieto-Vila M, Yan T, Calle AS, et al. iPSC-derived cancer stem cells provide a model of tumor vasculature. Am J Cancer Res. 2016;6(9):1906–21. Published 2016 Sep 1
Prince AM. An antigen detected in the blood during the incubation period of serum hepatitis. Proc Natl Acad Sci U S A. 1968;60:814–21. https://doi.org/10.1073/pnas.60.3.814.
Ramalho-Santos M, Willenbring H. On the origin of the term "stem cell". Cell Stem Cell. 2007;1: 35–8.
Rangarajan A, Hong SJ, Gifford A, Weinberg RA. Species–and cell type-specific requirements for cellular transformation. Cancer Cell. 2004;6:171–83.
Ribatti D. William Harvey and the discovery of the circulation of the blood. J Angiogenes Res. 2009 Sep;21(1):3.
Roberts SA, Gordenin DA. Hypermutation in human cancer genomes: footprints and mechanisms. Nat Rev Cancer. 2014;14:786–800.
Rodriguez R, Rubio R, Masip M, Catalina P, Nieto A, et al. Loss of p53 induces tumorigenesis in p21-deficient mesenchymal stem cells. Neoplasia. 2009;11:397–407.
Rous P. Transmissible avian neoplasm (sarcoma of the common fowl). J Exp Med. 1910;12:696–705. Transmission of a malignant new growth by means of a cell-free filtrate. Journal of Experimental Medicine 13, 397-411 (1911)
Shemer G, Podbilewicz B. The story of cell fusion: big lessons from little worms. BioEssays. 2003;25:672–82.
Sherley JL. Asymmetric cell kinetics genes: the key to expansion of adult stem cells in culture. Stem Cells. 2002;20:561–72.
Sonnenschein C, Soto AM. The Society of Cells: cancer and control of cell proliferation. New York, NY: Springer Verlag; 1999.
Sonnenschein C, Soto AM. Somatic mutation theory of carcinogenesis: why it should be dropped and replaced. Mol Carcinog. 2000 Dec;29(4):205–11.
Suy R, Thomis S, Fourneau I. The discovery of the lymphatic system in the seventeenth century. Part II: the discovery of Chyle vessels. Acta Chir Belg. 2016 Oct;116(5):329–35.
Takahashi K, Tanabe K, Ohnuki M, Narita M, Ichisaka T, Tomoda K, Yamanaka S. Induction of pluripotent stem cells from adult human fibroblasts by defined factors. Cell. 2007 Nov 30;131 (5):861–72.
Takahashi K, Yamanaka S. Induction of pluripotent stem cells from mouse embryonic and adult fibroblast cultures by defined factors. Cell. 2006 Aug 25;126(4):663–76.
Thankamony AP, Saxena K, Murali R, Jolly MK, Nair R. Cancer stem cell plasticity–a deadly Deal. Front Mol Biosci. 2020 Apr 30;7:79.
Thomson JA, Itskovitz-Eldor J, Shapiro SS, Waknitz MA, Swiergiel JJ, Marshall VS, Jones JM. Embryonic stem cell lines derived from human blastocysts. Science. 1998 Nov 6;282 (5391):1145–7.
Tyzzer EE. Tumor immunity. J Cancer Res. 1916;1:125–55.

Vaux DL. In defense of the somatic mutation theory of cancer. BioEssays. 2011;3:341–3.

Whitman RC. Somatic mutation as a factor in the production of cancer. A critical review of v. Hansemann's theory of anaplasia in the light of modern knowledge of genetics. J Cancer Res. 1919;4:181–202.

Wicha MS, Liu S, Dontu G. Cancer stem cells: an old idea—a paradigm shift. Cancer Res. 2006;66: 1883–90. https://doi.org/10.1158/0008-5472.CAN-05-3153.

Widschwendter M, Fiegl H, Egle D, Mueller-Holzner E, Spizzo G, et al. Epigenetic stem cell signature in cancer. Nat Genet. 2007;39:157–8.

Wunderlich V. Early references to the mutational origin of cancer. Int J Epidemiol. 2007;36:246–7.

Xing F, Okuda H, Watabe M, Kobayashi A, Pai SK, et al. Hypoxia-induced Jagged2 promotes breast cancer metastasis and self-renewal of cancer stem-like cells. Oncogene. 2011;30:4075–86.

Yamagiwa K, Ichikawa K. Experimentelle Studie über die Pathogenese der Epithelialgeschwülste mitt med fac Kaiserl univ tokyo. 1915:15.

Yan T, Mizutani A, Chen L, Takaki M, Hiramoto Y, Matsuda S, Shigehiro T, Kasai T, Kudoh T, Murakami H, Masuda J, Hendrix MJ, Strizzi L, Salomon DS, Fu L, Seno M. Characterization of cancer stem-like cells derived from mouse induced pluripotent stem cells transformed by tumor-derived extracellular vesicles. J Cancer. 2014 Jul 5;5(7):572–84.

Yan GN, Yang L, Lv YF, Shi Y, Shen LL, et al. Endothelial cells promote stem-like phenotype of glioma cells through activating the hedgehog pathway. J Pathol. 2014;234:11–22.

Yanger K, Zong Y, Maggs LR, Shapira SN, Maddipati R, et al. Robust cellular reprogramming occurs spontaneously during liver regeneration. Genes Dev. 2013;27:719–24.

Yimlamai D, Christodoulou C, Galli GG, Yanger K, Pepe-Mooney B, et al. Hippo pathway activity influences liver cell fate. Cell. 2014;157:1324–38.

Chapter 2
Culture of Cells: Basic Principles

Abstract The basic techniques for cell culture are introduced in this chapter. Primarily, the requirements for cell culture are an air-conditioned room of cleanroom class 10,000 (FS209E) equipped with lights and ventilation, a laminar flow hood, a CO_2 incubator, sterilizers with both saturated steam and dry heat, a low speed centrifuge with refrigerator, upright and inverted phase contrast microscopes, a freezer at −20, a deep freezer at − 80°C, a refrigerator at 4°C, a sink, disposable sterile plasticware (flasks, dishes, tubes, and pipettes), balance, ultrapure water, and a supply of media, and other reagents needed for the cell environment. Further helpful apparatuses are a pH meter, a cell counter (hemocytometer), a vacuum pump, a pipette-aid, micropipettes, a liquid nitrogen tank, a fluorescent microscope, and so on. Attention to safety and the maintenance of equipments is essential to understand the significance, the reasons, and the mechanisms. Contamination of microbials such as bacteria, yeast, staphylococcus, fungus, and mycoplasma should be strictly avoided in cell culture. Simultaneously careful attention not to overgrow but grow with sufficient cell density and to avoid passages for a long time because the phenotype of the cell may subject to change. This knowledge will help researchers with even a little prior experience to set up a suitable laboratory for basic cell culture.

Keywords Cell culture · Techniques · Cells · Principles

2.1 Introduction

By the end of the twentieth century, embryology, developmental biology, and cancer research had influenced the development of cell culture. At the beginning of the twentieth century, Ross Harrison (1870–1959) and Leo Loeb (1869–1959) used tissue and organ pieces in their experiments. They kept them alive with blood clots, salt solutions, and agar solutions in test tubes (Verma et al. 2020). One of the other pioneers in this field was Alexis Carrel (1873–1944). He adapted Harrison's method to culture tissue fragments from chicks and mice (Carrel and Burrows 1911a, 1911b). Facilities vary depending on how cell culture procedures will be applied to the individual investigation (Geraghty 2014). Although some space and equipment are required for routine operations in cell culture, the minimum could be

S. M. Afify, M. Seno, *Methods in Cancer Stem Cell Biology*,
https://doi.org/10.1007/978-981-99-1331-2_2

spared where cellular production and support are only minor responsibilities. The more active role is expected, the more elaborate facilities will be required. Therefore, some facilities could be combined into one laboratory or other separated into laboratories with each single specific function. There are some equipments essential and common to all laboratories, regardless with scale of the experiment. For instance, microscopes, pH meter, centrifuges, etc., should be essential. Central services such as animal handling facilities, and glassware washing facilities are preferrable to be ready-to-use. Also, the essential equipment, for example, microscopes, pH meter, centrifuges, etc., should be considered. Cell culture techniques for cancer biology laboratories will be discussed in the notes integrated in this chapter. The other information will be presented in the same manner. Although this chapter focuses on the culture of mammalian cells and cancer stem cells (Afify et al. 2019), the topics discussed here are generally applicable to all types of cell culture.

2.2 Basic Requirements of Cells in Culture

All equipment must be ready in optimal working condition before you start culture cells. Cell culture facility requires essential conditions such as an air-conditioned room of cleanroom class 10,000 (FS209E) equipped with lights and ventilation, a laminar flow hood, a CO2 incubator, sterilizers both with saturated steam and dry heat, a centrifuge at a low speed with refrigerator, upright and inverted phase contrast microscopes, a freezer at −20 and a deep freezer at − 80 °C, refrigerators at 4 °C, sinks, disposable sterile plasticware (flasks, dishes, tubes, pipettes), balances, ultrapure water, and media and other reagents (Fig. 2.1).

2.2.1 Equipments

2.2.1.1 Laminar Flow Hood

Laminar flow hoods, or biosafety cabinets, are probably the most important parts of equipments for cell culture because, they provide filtered air around the work surface when installed properly. Depending on the research requirements laminar flow hoods (cabinets) are categorized into three types

A. Class I,
B. Class II
C. Class III

Biosafety Cabinet Class I: This type applies to laboratory settings in which a work handles with low-hazardous-risk microbes that pose little toor no infection, and agents of no or minimal potential hazard. It is recommended for proper

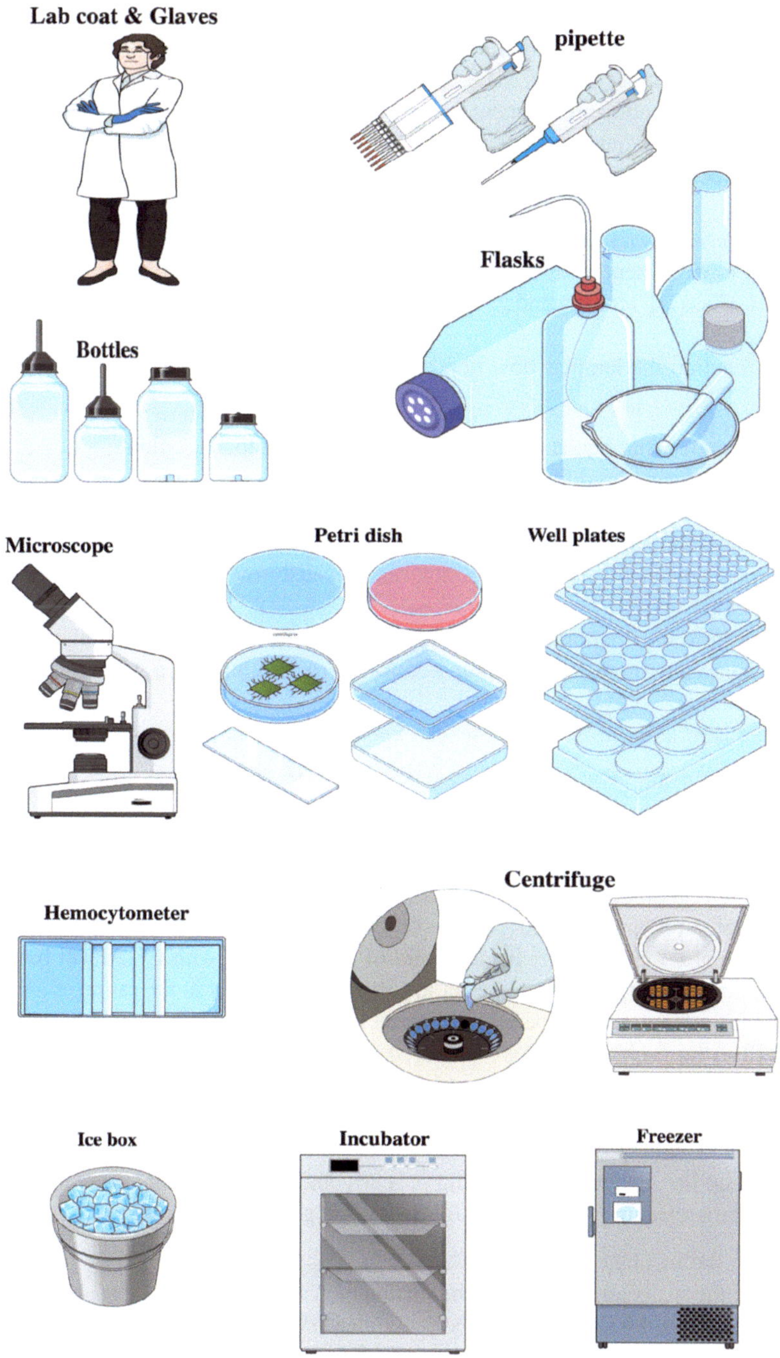

Fig. 2.1 Representative equipments essential for cell culture

microbiological techniques, which supplies essential levels of protection of workers and environment from pollution. But it does not protect cultures from contamination.

Biosafety Cabinet Class II: This type is designed far more restrictive than Class I to allow an aseptic environment essential for experiments of cell culture. It is recommended for handling exclusively harmful materials, e.g., pathogen cultures, virally infected cells, organisms labeled with radioisotopes, and carcinogen or toxic reagents.

Biosafety Cabinet Class III: This type is designed gas-tight, to supply the workers and the environment with the highest achievable level of protection. A biosafety cabinet Class is recommended for the work that handles seriously pathogenic microbes.

To function the laminar flow hood properly, the following guidelines should be kept:

- The use of clean bench should strictly be restricted to cell culture.
- Work surfaces should be sterilized after any work.
- All contaminated liquid or solid wastes should be disposed after sterilized decontaminated.
- Prepare mechanical pipetting devices because mouth pipetting should be forbidden.
- Wear individual laboratory coats, gowns, or uniforms to prevent contamination.
- Change your shoes to those prepared for the cell culture area when entering the laboratory.
- Keep laboratory doors closed when experiments are in progress.

2.2.1.2 Cell Culture Incubators

Cells require a precisely controlled environment to grow. Cell culture incubators are capable of supplying suitable growth conditions, such as control of vaporizing with humidity, of temperature by heating, and of pH of media with CO2 level balanced in routine. Generally, they can be set to run at temperatures in the range of 28 °C to 37 °C depending on the type of culture and set to provide CO2 level approximately 5 to 10%. Some incubators are prepared to control three gasses of NO2 and O2 levels as well as CO2. In order to reduce the risk of microbial contamination within the incubator, copper is used in the panels inside due to the inhibitory effect on microbial growth.

To keep the incubator functioning properly and to protect cells from contamination, regular cleaning is necessary following the guidelines below:

- Clean the incubator every two weeks.
- Wipe the walls and shelves in the incubator with 70% ethanol.
- After drying the shelves could be put under UV light for at least 5 h.
- Fresh, sterile, and distilled water should be added into the container.

2.2.1.3 Cleaning and Sterilization Facilities

In a cell culture laboratory, the most significant aspect is clean and sterile, regardless of the availability of sterilization, disposable labware media and reagents. A laboratory must at least keep clean and sterilize glassware from time to time. Researchers should be trained in proper handling of glassware that comes into direct contact with cell cultures and is a potential cause of contamination.

The daily procedures of cleaning and sterilizing glassware should carefully be performed so as to avoid problems such as microbial contamination, residual detergent after washes, etc.

Aseptic and preparation areas should preferably be separated from the cleaning facilities. If the space is limited to separate the areas, they may be combined if contamination with microorganisms is not experienced. Since cell culture procedures usually require a large number of pipettes, use of drying oven is convenient to get most glassware dried after wash (Fig. 2.1).

2.2.2 *Reagent Essential for Cell Culture*

2.2.2.1 Media

Several conventional media are available for the growth of established cell lines raised below. Some of them are supplemented with serum while the others are chemically defined without serum. Sodium bicarbonate is usually used in the media to adjust pH at neutral, which is often monitored with visible color of Phenol red, under the buffering condition with an appropriate percent of CO2.

Eagle Media

Basal Medium Eagle.

Synthetic media developed in the 1950s are used today. It contains amino acids, carbohydrates, vitamins, and salts. For growth, serum or other components are added to the basal media, such as growth factors.

Various kinds of media have different compositions with the most essential ingredients. Basal medium Eagle (BME) was the first media among those developed. Based on the study by Harry Eagle BME is capable to support a wide variety of malignant cells as well as normal ones (Eagle 1955).

A. *Eagle's Minimum Essential Medium (MEM)*

According to cellular requirements, Eagle modified BME to a twofold concentration of the majority of amino acids as minimum essential Medium (MEM). MEM formula does not contain non-essential amino acids, which cells can biosynthesize, but researchers can add them to reduce the biosynthesis load (Eagle 1959).

B. *Dulbecco's Modified MEM (DMEM)*

To study how polyoma virus forms plaques in mouse embryonic cells, Dulbecco and Freeman developed a modified MEM as DMEM that contained four fold concentration of the amino acids and vitamins in BME. Since then, many modifications have been made in DMEM, including the supplement of iron, pyruvate and non-essential amino acids, such as glycine and serine. It is also possible to increase glucose concentrations to meet the nutritional requirements of cells (Dulbecco and Freeman 1959).

Ham Media

A. *Ham's F-10*

Richard G. Ham developed a synthetic cell culture media supplemented with two serum derived proteins, serum albumin and fetuin as a nutrient mixture of F10, which is serum-free condition (Ham 1962). He found a single Chinese hamster ovary (CHO) cell was capable to form colonies, Copper and zinc were included in this medium for the first time in Ham's F-10 medium (Ham 1963).

B. *Ham's F-12*

Ham's F10 medium was improved to Ham's F12 medium, the serum albumin and fetuin being replaced by atty acid and polyamine, linoleic acid and putrescine. Although the contents of potassium phosphate and vitamin except for choline and inositol are reduced from F10, F12 medium is considered as the first chemically defined medium with amino acid contents more than those in F10 (Ham 1965).

Roswell Park Memorial Institute (RPMI) medium

TMcCoy's 5A medium (McCoy et al. 1959) which was originally developed for sarcoma culture in vitro, was modified as RPMI medium for the long-term culture of leukemic cells and peripheral blood lymphocytes (Moore et al. 1966). RPMI medium contains high concentration of phosphate and low concentrations of calcium and magnesium. This medium has different variations distinguished by numbers while 1640 is the most popular medium for suspension cultures such as those of white blood cells, lymphocytes, and hybridomas (Moore et al. 1966)

2.2.2.2 Serum

Serum was first used as a cell culture component together with balanced salt solution, embryonic tissue extract, minerals in varying combination (Sanford et al. 1948; Scherer et al. 1953). While the components in serum have been extensively studied and analyzed to identify most of all the nutrients essential for growth, serum is still used as an efficient source of numerous essential elements growth. In many cases cells can survive and grow in serum containing culture medium in a manner similar to those counterparts in vivo. Fetal bovine and newborn calf (and sometimes human or equine) are the most popular origin of serum for cell culture. Serum is usually used between 5 and 20% depending on the type of cells.

2.2.3 Basic Techniques of Cell Culture

2.2.3.1 Cell Culture

A cell culture is a technique that involves the cultivation of cells outside of living organisms under controlled conditions (e.g., temperature, pH, nutrients, etc.). To obtain cells, one can either obtain them from a cell bank or isolate them from tissue (Fig. 2.2). Cultures from cell banks require reviving/thawing prior to beginning.

2.2.3.2 Reviving the Cell

To initiate a cell culture, frozen cryopreserved cells must be thawed.

Wake up the cell should follow next steps:

1. Pick up a vial of frozen cell stocks from a LN2-tank.
2. Soak the vial in a 37 °C water bath to thaw the cells.
3. Before ice is almost melted, sterile the vial wiping with 70% ethanol cotton before open the vial.
4. Transfer the vial to clean a bench.

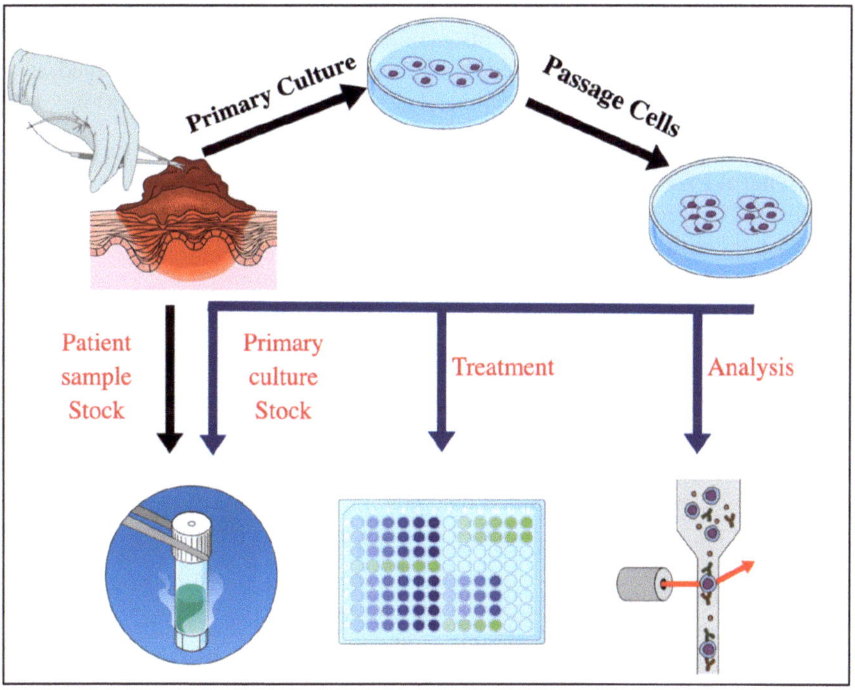

Fig. 2.2 A schematic representation of the basic methods in cell culture

5. Dilute the cryoprotectant liquid with a pre-warmed medium and mix the cells with gentle tapping.
6. Centrifuge the cells at 180 g at 25°C for 5 min.
7. Discard the supernatant carefully using a sterile Pasteur pipet, without disrupting the pellet.
8. Suspend the cells with prewarmed fresh medium.
9. Wash the pellet with 1X PBS or fresh medium.
10. Centrifuge the cells at 180 g 25°C for 5 min.
11. Discard the supernatant carefully using a sterile Pasteur pipet without disrupting the pellet.
12. Suspend the cells with prewarmed fresh medium.
13. Seed the cells on three 60-mm dishes, with different cell densities of low and high.
14. Incubate the dishes in a humidified incubator at 37°C conditioned with CO2.

2.2.3.3 Passage

When the cells are adhesive to the surface of culture vessel, they need to be detached. There are several proteases available for the purpose. Typically, collagenase, dispase, and trypsin are used. Since trypsin is inhibited by calcium and magnesium ions, it is necessary to wash the cell layer with PBS containing EDTA in advance. To stop the reaction, addition of medium containing serum is effective because serum contains trypsin inhibitor. In contrast, collagenase and dispase are active when calcium ions are present. Where possible, cancer cells should be passaged at an 80–90% confluency. It is not recommended to allow them to reach 100% confluence or overgrowth.

The General Procedure Is as Follows

1. Discard culture medium.
2. Wash the dish with PBS containing 0.01% EDTA.
3. Disperse cells with enzyme solution (0.025% Trypsin).
4. Incubate the dish in 37 °C to promote the enzymatic reaction.
5. Examine the detachment of the cells under a microscope.
6. Detach cells by mild mechanical disturbances.
7. Add a pre-warmed medium containing serum to stop the reaction.
8. Dissociate the cells into single-cell suspensions by pipetting the medium several times.
9. Centrifuge the cell suspention at 180 g at 25°C for 5 min.
10. Dispose the supernatant carefully with a sterile Pasteur pipet, without disrupting the pellet.
11. Tap the tube to loosen the cell pellet.
12. Add fresh medium to the cells.
13. Pipette up and down to resuspend the cells with a sterile Pasteur pipet.
14. Count cells using a microscope with hemocytometer or an automated cell counter.

15. Seed cells at 0.3 × 10^6/60-mm dish.
16. Incubate at 37 °C in a humidified CO_2 incubator.

2.2.3.4 Preparation of frozen cell storage

It is important to cryopreserve aliquots of the culture once the optimal growth parameters have been established for a particular cell line.

The next steps should be followed to store the cells as frozen state:

1. Prepare the cells to be stored at 70 to 80% confluent to keep high reviving efficiency when reviving the cells, then take aliquots of cells for stock.
2. Centrifuged at 1000 rpm for 5 min.
3. Discard the supernatant.
4. Add cryoprotectant medium to the cell pellet. Typically, DMEM high glucose containing 50% FBS and 10% DMSO is used. If DMSO is not recommended, glycerol may be replaceable. Some cryoprotectant media without DMSO, such as CELLBANKER®, Cell Reservoir One, etc., are commercially available.
5. Adjust the count of the cells to be 1× 10*6 cells/vial.
6. Each vail contains 250 ul of the cell reservoir.
7. Close the vail and name the cells.
8. Put the cryotubes in a container made of foamed styrol pre-cooled at 4°C. Note: The container should be able to tightly package the tubes. Some containers, such as CoolCell® and BICELL, etc., are commercially available for packaging. Then put the containers into a -80°C immediately and keep one overnight.
9. Transfer the tubes to liquid nitrogen tank. The cells should be stored at the gas phase of N2.

References

Afify SM, Chen L, Yan T, Calle AS, Nair N, Murakami C, Zahra MH, Okada N, Iwasaki Y, Seno A, Seno M. Method to convert stem cells into cancer stem cells. Methods Protoc. 2019 Aug 16;2 (3):71.

Carrel A, Burrows MT. An addition to the technique of the cultivation of tissues in vitro. J Exp Med. 1911a;14(3):244–247.2.

Carrel A, Burrows MT. Cultivation of tissues in vitro and its technique. J Exp Med. 1911b;13(3): 387–96.

Dulbecco R, Freeman G. Plaque production by the polyoma virus. Virology. 1959;8:396–7.

Eagle H. The specific amino acid requirements of a human carcinoma cell (strain HeLa) in tissue culture. J Exp Med. 1955;102:37–4.

Eagle H. Amino acid metabolism in mammalian cell cultures. Science. 1959;130:432–7.

Geraghty RJ, Capes-Davis A, Davis JM, Downward J, Freshney RI, Knezevic I, Lovell-Badge R, Masters JR, Meredith J, Stacey GN, Thraves P, Vias M, Cancer Research UK. Guidelines for the use of cell lines in biomedical research. Br J Cancer. 2014 Sep 9;111(6):1021–46.

Ham RG. Clonal growth of diploid Chinese hamster cells in a synthetic medium supplemented with purified protein fractions. Exp Cell Res. 1962;28:489–500.

Ham RG. Albumin replacement by fatty acids in clonal growth of mammalian cells. Science. 1963;140:802–3.

Ham RG. Clonal growth of mammalian cells in a chemically defined, synthetic medium. Proc Natl Acad Sci U S A. 1965;53:288–93.

Mccoy TA, Maxwell M, Kruse PF Jr. The amino acid requirements of the Jensen sarcoma in vitro. Cancer Res. 1959;19(6, Part 1):591–5. PMID: 13671417

Moore GE, Ito E, Ulrich K, Sandberg AA. Culture of human leukemia cells. Cancer. 1966;19:713–23.

Sanford KK, Earle WR, Likely GD. The growth in vitro of single isolated tissue cells. J Natl Cancer Inst. 1948;9(3):229–46. PMID: 18105872

Scherer WF, Syverton JT, Gey GO. Studies on the propagation in vitro of poliomyelitis viruses. IV. Viral multiplication in a stable strain of human malignant epithelial cells (strain HeLa) derived from an epidermoid carcinoma of the cervix. J Exp Med. 1953;97(5):695–710. https://doi.org/10.1084/jem.97.5.695. PMID: 13052828; PMCID: PMC2136303

Verma A, Verma M, Singh A. Animal tissue culture principles and applications. Anim Biotechnol. 2020;269–93

Chapter 3
Stem Cell Culture from Embryos

Abstract Embryonal carcinoma cells were firstly used as the in vitro model for early mouse development because the morphology and pluripotency are similar with embryonal cells in 1964. These cells were isolated from the teratocarcinoma in vitro as a stem cell even with genetic aberrations. Then two independent groups succeeded in culturing the ESC from the inner cell mass of mouse embryo in 1981.

Embryonic stem cells (ESCs) are pluripotent stem cells which will give rise to all the components of somatic cells in the whole body when embryos develop. Refraining the differentiation, ESCs will keep self-renewing to provide enough numbers of ESCs available for tissue regeneration. Therefore, ESCs should be valuable materials to understand the mechanisms of the development of specific organ structures with tissue-specific cells (Fig. 3.1). Many protocols to culture ESCs could be found as the standards of mammalian stem cell culture. However, the maintenance of ESCs in undifferentiated condition has been a big issue in the developmental cell biology because additional considerations have been required to maintain their characteristics of self-renewal and pluripotency along with the passages. This chapter describes how to culture and passage ESCs with/without feeder cells maintaining their unique properties.

Keywords Embryonic stem cells · Self-renewal

3.1 Introduction

Embryonic stem cells (ESCs) were first isolated in the 1980s by several independent groups (Vans and Kaufman 1981; Axelrod 1984; Wobus et al. 1984; Doetschman et al. 1985). These investigators found the pluripotent characteristics of ESCs to differentiate into any phenotypes derived from the three primary germ lineages of endoderm, mesoderm, and ectoderm. Gossler et al. described the advantageous application of ESCs to establish transgenic animals (Gossler et al. 1986). The isolation of ESCs from primate as well as human is one of the current foci in the stem cell technology (Thomson et al. 1995; Thomson et al. 1998; Shamblott et al. 1998; Reubinoff et al. 2000).

S. M. Afify, M. Seno, *Methods in Cancer Stem Cell Biology*,
https://doi.org/10.1007/978-981-99-1331-2_3

Basically, ESCs are isolated from early-stage embryos maintaining the pluripotency. Although a fertilized egg is essential to prepare ESCs, cells are generally harvested from the blastocyst, which is 3–5 days of post-fertilization. The outer cell layer of the blastocyst, called the trophoblast, contains a fluid-filled cavity, the blastocoele, and an inner cell mass of 10–20 cells. The inner mass of cells, which compose so-called embryoblast, are dissected for culture. Occasionally, the ESCs may be obtained from the stage before the formation of the blastocyst, which is the so-called morula formed at approximately 80 h after fertilization.

The isolation and maintenance of ESCs faces several challenges due to the naïve and prime characters of stemness including pluripotency, which are affected by the epigenetic effects and chromosomal changes during culture. Protocols are continuously improved and vary depending on the stages of ESCs. After being isolated from the blastocyst in the inner cell mass (ICM), ESCs will continue to indefinitely grow in an undifferentiated state if maintained in optimal conditions. Since ESCs are sensitive to pH and temperature changes, overgrowth, daily care for these cells in culture dishes is important and imperative. ESCs will spontaneously differentiate without sufficient maintenance, even in the presence of feeder layers and leukemia inhibitory factor (LIF). In addition, ESCs growing in log phase are critically optimal for transformation with foreign gene introduction. Many opportunities exist in stem cell biology. The process of developing stem cell lines will be extended to develop new technologies, which explore further stages of advancement in cell engineering. Therefore, this chapter will present the methods used in our laboratory to isolate ESCs and maintain them undifferentiated in good conditions for further use (Fig. 3.1).

3.2 Materials

3.2.1 Equipment

- Corning® Matrigel®, growth factor reduced, phenol red-free (BD, cat. no. 356231).
- Dulbecco's modified Eagle's medium-high glucose (Wako, Osaka, Japan (catalogue number: 044–29,765)).
- Trypsin-EDTA (0.25%) (Nacalai Tesque, Kyoto, Japan, Cat. No: 327777–44).
- Fetal bovine serum (FBS, Gibco, Life Technologies, Massachusetts, USA, Cat. No: 10437–028).
- Penicillin/streptomycin mixed solution (100 U/mL) Nacalai Tesque, Kyoto, Japan, Cat. No-26253-84).
- 70% ethanol (Sigma-Aldrich; Cat. No.: 459836–2).
- Liquid N_2 storage tank.
- Hank's balanced salt solution (HBSS) Genesee Scientific, El Cajon, USA.
- Endothelial basal medium EBM2 media (EBM-2 SingleQuots Kit, Lonza, Switzerland).

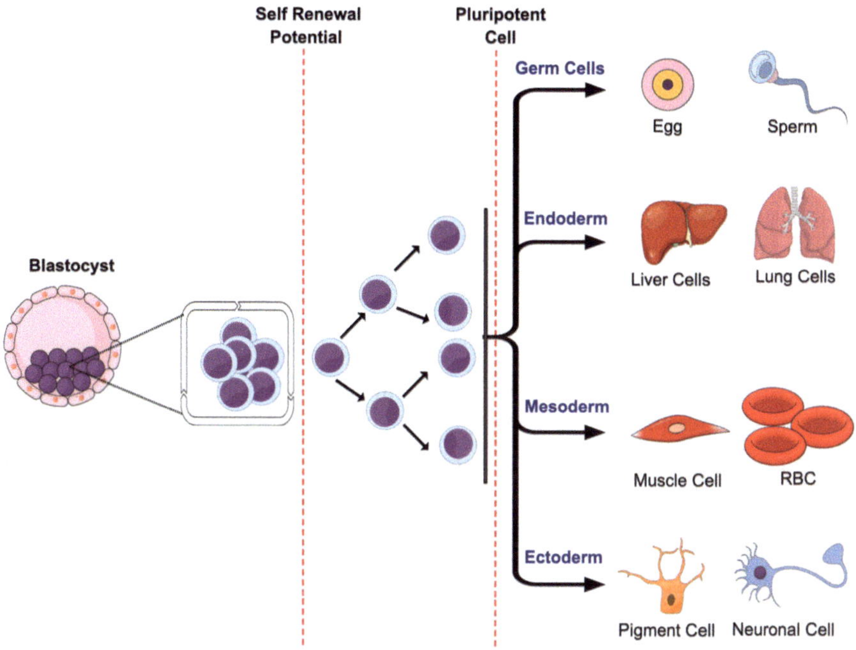

Fig. 3.1 Represented scheme for differentiation potential of stem cells into different types of cells

3.2.2 Equipment

- Eppendorf Centrifuge 5415R, Eppendorf AG, 22331 Hamburg, Germany.
- Sanyo MCO-19AIC (UV) CO2 Incubator, Marshall Scientific, Hampton, USA.
- Type A2 Biological Safety Cabinets (E-Series).
- Olympus IX81 microscope (Olympus, Tokyo, Japan).
- Tissue culture-treated plate, 60 mm dish; TPP, AG Schaffhausen, Switzerland, Cat. No 93060.
- Tissue culture-treated plate, 100 mm dish; TPP, AG Schaffhausen, Switzerland, Cat. No 93100.
- Filter max 250 mL, TPP, Switzerland, Cat. No 99255.
- Falcon® Conical Centrifuge Tubes (15 mL; BD Falcon, New York, USA, Cat. No 352095).
- Falcon® Conical Centrifuge Tubes (50 mL; BD Falcon, New York, USA, Cat. No 352070).
- 37 °C water bath.

3.2.3 *Reagent Preparation*

Stem Cell Medium		
Mix the following for 500 mL		
Dulbecco's modified Eagle's medium-high glucose		412.5 mL
FBS		75 mL
Penicillin/streptomycin mixed solution	(50 U/mL)	2.5 mL
Gibco™ L-glutamine (200 mM)	(2 mM)	5 mL
MEM non-essential amino acids solution	(1 mM)	5 mL

0.1%(w/v) Gelatin
Dissolve 0.5 g of gelatin (from porcine skin) in 500 ml distilled water and autoclave. Store at room temperature indefinitely.

Mouse Embryonic Fibroblasts (MEFs) Medium
To prepare 500 mL of complete MEF medium, aseptically mix the following components:

Component	Volume
DMEM	445 mL
FBS, ESC-qualified	50 mL
MEM non-essential amino acids solution, 10 mM	5 mL
β-Mercaptoethanol, 1000X	500 μL

Complete MEF medium can be stored at 2–8 °C for up to 1 week.

Prepare PBS
Phosphate-buffered saline solution (PBS): Dissolve the following salts in 1000 ml of deionized water: 8 g NaCl, 0.2 g potassium chloride (KCl), 1.44 g disodium phosphate (Na2HPO4*2H2O), 0.2 g monopotassium phosphate (KH2PO4). Adjust pH to 7.4 using sodium hydroxide (NaOH) or hydrogen chloride (HCl).

3.3 Methods

In the current chapter we tried to evaluate the step-by-step procedure for isolating ESCs from embryos (Fig. 3.2). First of all, a blastocyst or earlier was obtained from the uterus. Then ESCs were cultured on the feeder layer of MEF cells, which were treated with mitomycin C or radiation, for 2 weeks. After 2 weeks very bright colonies of ESCs appeared. Stable colonies were moved to feeder-less culture condition and kept the phenotype undifferentiated without MEF cells. All materials, glassware, plastic, etc., have to be sterile and all procedures should be carried out under aseptic conditions with a laminar flow cabinet (biological hazard standard).

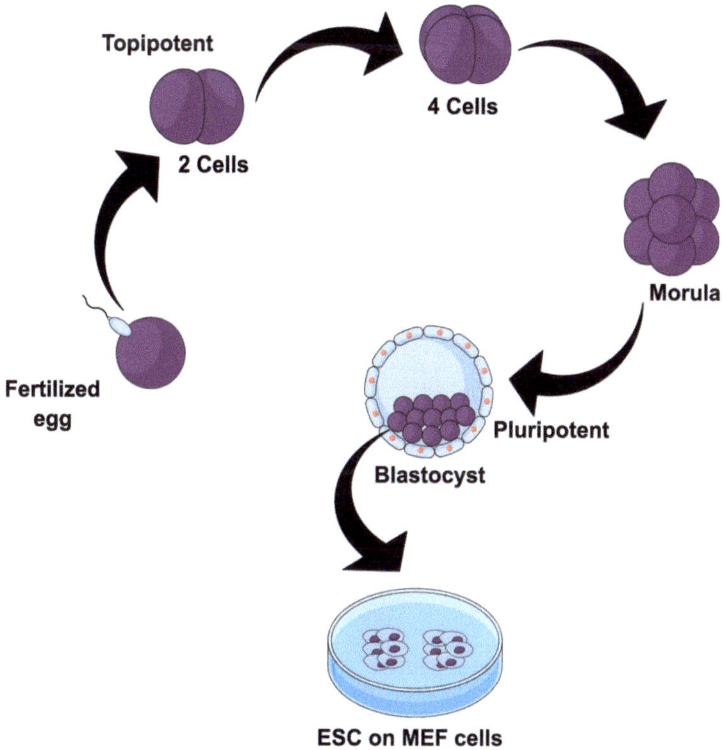

Fig. 3.2 Blastocyst development and isolating ESC

3.3.1 Preparation of Gelatin-Coated Dishes

1. Pick 0.2% gelatin solution up in the laminar flow cabinet from 4 °C.
2. Pour enough volume of 0.2% gelatin solution with a pipette to cover the bottom of the dish. Note 1.
3. Let the dish stand for 30 min at 37 °C in a 5% CO_2 incubator.
4. Aspirate the excess gelatin solution. Note 2.

3.3.2 Preparation of MEF-Coated Dishes

1. Pick up the vial of mitotically inactivated MEF from liquid nitrogen storage with metal forceps and put on ice.
2. Immerse the vial in a 37 °C water bath without submerging the cap. Note 3.
3. Pick up the vial from the water bath when the frozen cells become dissolved with just a few ice crystals.
4. Spray the outside of the vial with 70% ethanol.

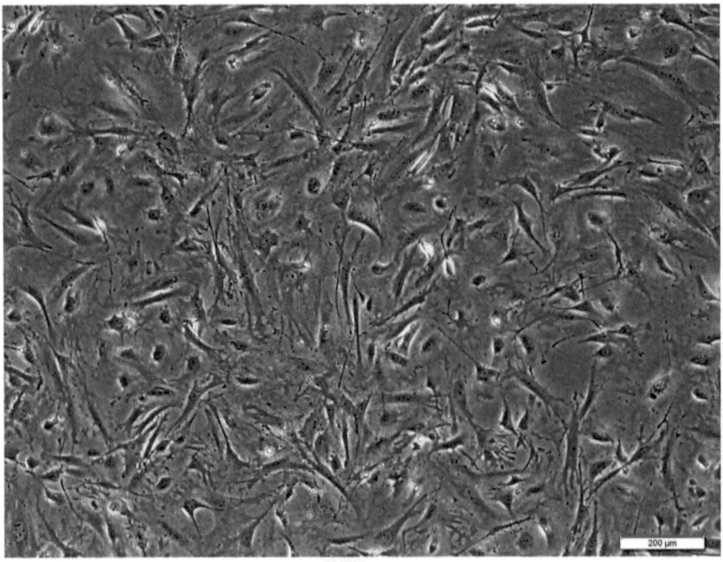

Fig. 3.3 Representative image for MEF. Scale bar = 200 μm

5. Transfer the vial to the clean bench.
6. Add 5 mL of pre-warmed MEF medium to a 15-mL sterile conical tube.
7. Take 1 ml of pre-warmed MEF medium to the cells in the vial.
8. Pipette the thawed cells gently and transfer the cells into the conical tube. Note 4.
9. Mix the cells gently by slow pipetting in the MEF medium.
10. Centrifuge the cells at 200 xg for 5 min at 25 °C.
11. Aspirate the supernatant without disturbing the pellet.
12. Resuspend the cell pellet in pre-warmed MEF medium.
13. Pick up 50 μL of the suspension and count the cells with trypan blue using a hemocytometer.
14. Add 4 mL of pre-warmed MEF medium into each gelatin-coated dish from Step 4 in the previous section.
15. Seed appropriate number of MEF at a density of 8×10^4 cells/cm^2 per dish, plate, or flask coated with gelatin.
16. Incubate the MEF into a 37 °C in 5% CO_2 incubator for more than 24 h. Note: MEF dishes are ready at the best condition 3–4 days after plating (Fig. 3.3) but available up to approx. 2 weeks in incubator.

3.3.3 Obtaining Blastocyst Stage Embryos

Blastocysts can be obtained from female mice 3 days after mating.

1. Place two female mice per male mice in a cage. Note 5.
2. Check for copulation plugs daily. Note 6.
3. Separate plugged females and label for blastocysts embryos 3 days later.
4. Fill a 1.5-ml microcentrifuge tube with pre-warmed M2 medium.
5. Sacrifice plugged females humanely by cervical dislocation.
6. Cut the skin and the peritoneum to view the abdominal organs.
7. Locate the ovary, oviduct, and uterus.
8. Dissect the reproductive organs such as ovary, oviduct, and two uterine horns in one intact piece. Note 7.
9. Collect and transfer the organs into the 1.5-ml microcentrifuge tube with M2 medium.
10. Place the tissues into a 60-mm culture dish filled with pre-warmed M2 medium.
11. Cut off the oviduct and the ovary under a dissecting microscope or binocular stereomicroscope.
12. Prepare a 27-gauge needle attached to a 1-ml syringe filled with pre-warmed M2 medium.
13. Insert the syringe into one uterine horn, and then flush out the embryos with 0.5 ml of M2 medium.
14. Flush again to release any remaining blastocysts.
15. Repeat Steps 12–14 with the other horn.
16. Flush blastocyst stage embryos from both uterine horns.
17. Transfer the embryos through several M2 drops to wash away uterine fluids and debris.
18. Transfer one washed embryo into a 4-well dish with fresh MEF feeder layer.
19. Incubate the dish at 37 °C in a 5% CO_2 incubator.

3.3.4 *Thawing and Plating ESCs*

1. Aspirate the MEF medium from a MEF dish from Sect. 3.2.
2. Add pre-warmed ESC culture medium to the dish 3–4 h before plating ESCs.
3. Pick up the vial of ESCs from liquid nitrogen storage with metal forceps.
4. Immerse the vial in a 37 °C water bath without submerging the cap.
5. Remove the vial from the water bath, when half of the vial is melted.
6. Sterilize the outside of the vial with 70% ethanol and place it in a laminar flow hood.
7. Remove cap and slowly add 1 ml of pre-warmed MEF Medium drop by drop.
8. Pipette the thawed cells gently into a sterile 15-mL conical tube.
9. Add 4 ml of ES culture medium to 15 mL conical tube.
10. Centrifuge the cells at 200 xg for 5 min.
11. Aspirate the supernatant without disturbing the cell pellet.
12. Resuspend the cells in a suitable volume of pre-warmed ES culture medium.
13. Take 50 μl of the suspension and mix with trypan blue and count the cells with a hemocytometer.

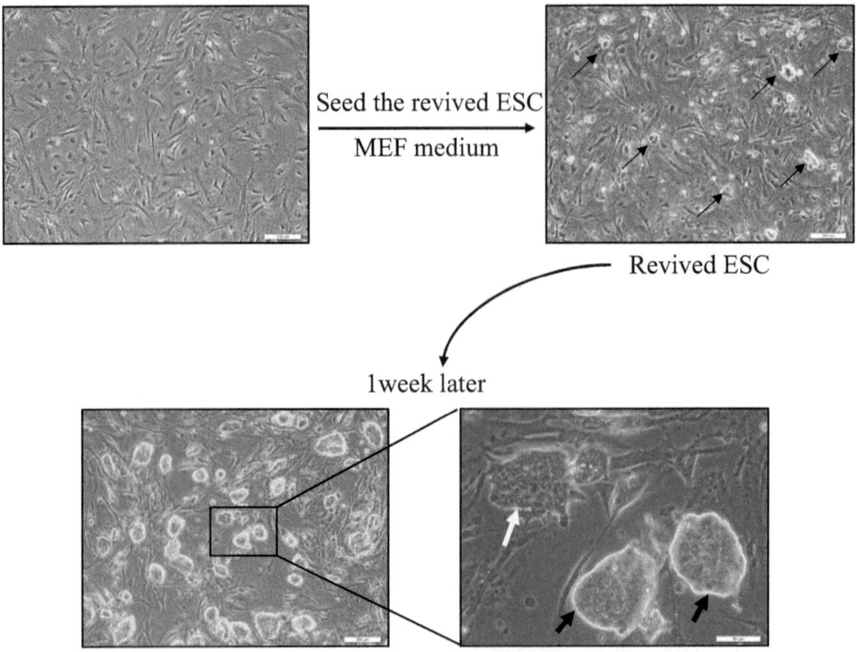

Fig. 3.4 Representative images of ESCs revived on MEF. Colonies of undifferentiated ESCs (black arrows) and those of differentiating ESCs (white arrow). Scale bars represent 200 μm (top, bottom left) and 50 μm (bottom right)

14. Seed ESCs onto the dish (from Step 2) at a plating density of roughly 5×10^4 cells/cm^2.
15. Incubate the ESC dishes at 37 °C in 5% CO_2 incubator.
16. Check the cells changing medium every two days up to 1 week (Fig. 3.4). Note 8.
17. Divide colonies of ESCs into at least five 60-mm dishes when the confluency reaches to 70–80%. Note 9.

3.3.5 *Splitting ESCs on MEF*

When ESCs become crowded with large colonies, a passage is recommended at this point.

1. Add fresh ES culture medium to ESC dishes.
2. Incubate the dishes at 37 °C in a 5% CO_2 incubator for 2–6 h before trypsinization.
3. Pick up the ESC dish from the incubator.
4. Rinse the ESCs with PBS at room temperature.

5. Add 0.25% trypsin solution just to cover the bottom of the dish.
5. Place the dish in 37 °C incubator until the cells begin to visibly detach from the bottom of the dish. Note 10.
6. Add equal volume of culture medium with trypsin solution to the dish to stop trypsinization.
7. Pipette up and down slowly to break down the colonies.
8. Transfer the cells into a sterile 15-ml conical tube.
9. Centrifuge gently to pellet ES cells at 200 xg for 5 min at 25 °C.
10. Aspirate the supernatant with a pipette attached to a vacuum trap.
11. Resuspend cells in 2 ml of pre-warmed ES cell medium.
12. Aspirate the medium from MEF dishes prepared earlier.
13. Add 4 mL pre-warmed medium.
14. Take 50 μl of the suspension and mix with trypan blue and count the cells with a hemocytometer.
15. Seed the ESCs onto the MEF dishes at a density of 0.3×10^6 cells/60-mm dish.
16. Incubate the plates at 37 °C with 5% CO_2.
17. Monitor the colonies and take photos periodically (Fig. 3.5).

3.3.6 Passage ESC on Feeder-Less Dish

1. Aspirate the medium from ESCs on the MEF dish at 70% confluent.
2. Wash the cells with 3 mL of PBS and aspirate. Repeat this step twice.
3. Add 500 μL dissociation buffer and spread well over the bottom.
4. Incubate the dishes at 37 °C in a 5% CO_2 incubator for 5 min.
5. Pick up the plates from the incubator.
6. Observe the cells under an inverted microscopy. Note 11.
7. Aspirate the dissociation buffer gently.
8. Add 2 mL of ES cell medium.
9. Detach the cells with a cell scraper or gentle pipetting.
10. Collect the colonies and transfer to a sterile 15-ml conical tubes.
11. Suspend the cells by pipetting gently so the colonies of ES cells are not completely destroyed.
12. Centrifuge at 100 xg for 5 min at 25 °C.
13. Aspirate the supernatant.
14. Resuspend the cells in 2 ml of pre-warmed ES cell medium.
15. Bring the gelatin or Matrigel-coated dishes from the incubator.
16. Aspirate the excess amount of gelatin or diluted Matrigel solution.
17. Add 4 mL pre-warmed medium.
18. Take 50 μl of the suspension and mix with trypan blue and count the cells with a hemocytometer.

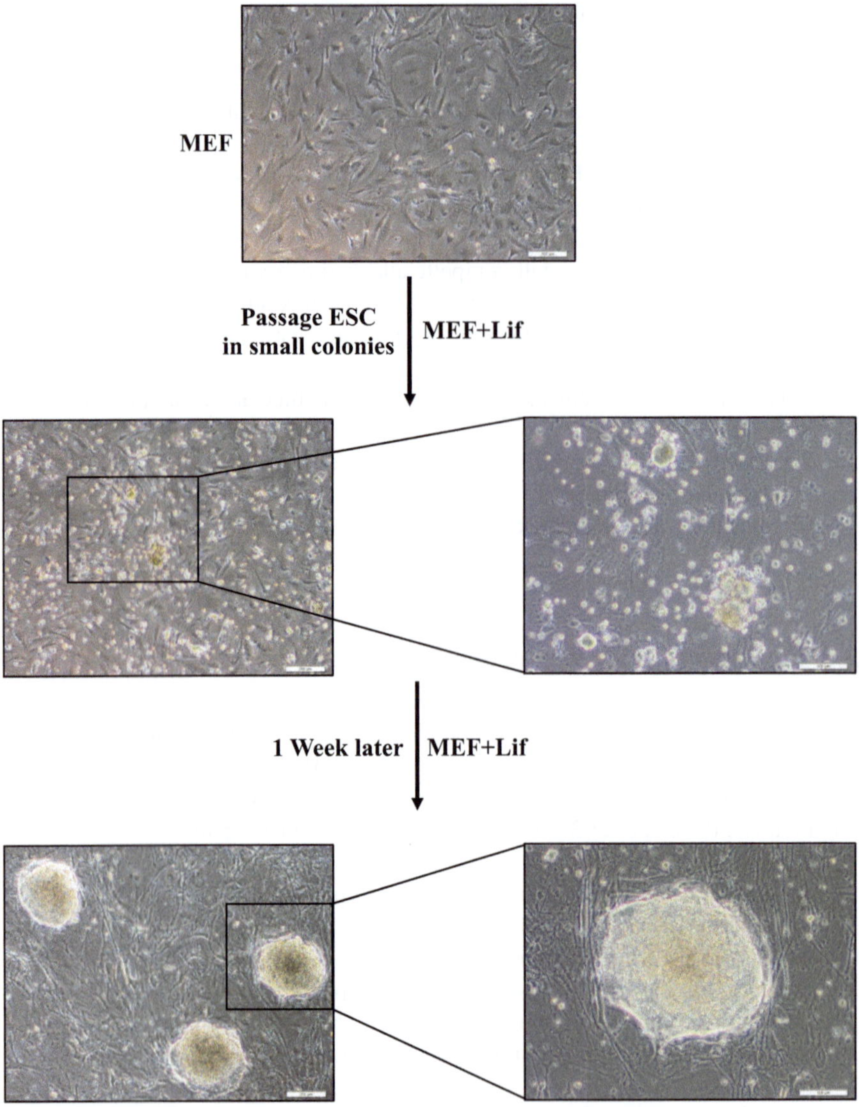

Fig. 3.5 Representative images of ESCs passaged on MEF. ESCs seeded on MEF (middle) formed colonies after 1 week (bottom). Scale bars represent 200 μm (left) and 50 μm (right)

19. Seed the ESCs onto the gelatin-coated dishes at a density of approx. 3 x 10^5 cells/60-mm dish.
20. Incubate the dishes at 37 °C in a 5% CO_2 incubator for around 1 week until ESC colonies become very bright and clear (Fig. 3.6). Note 12.

Fig. 3.6 Representative images of ESCs passaged on gelatin-coated dishes. Scale bars represent 200 μm (top, bottom left) and 100 μm (bottom right)

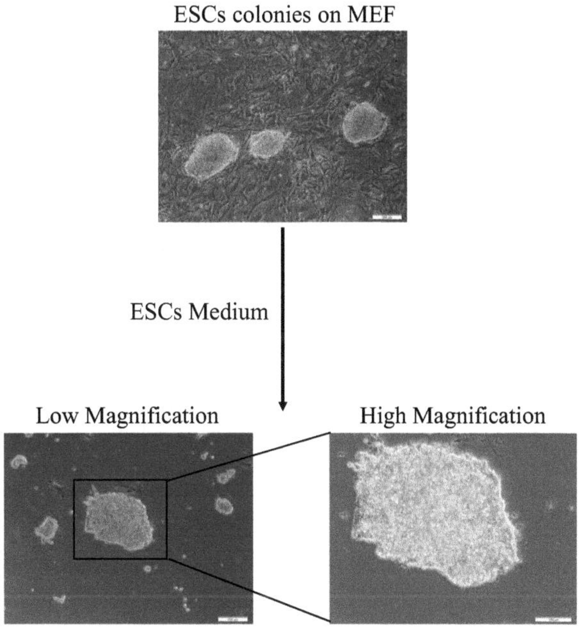

3.4 Notes

1. Approx. 2 mL for a 60-mm dish and 5 mL for a 100-mm dish.
2. The dishes are ready for immediate use. Do not try to store the dishes.
3. Be careful not to get water into the vial by soaking it above the bottom line of cap. There is a high risk of contamination in this step.
4. This step should be done as quick as possible.
5. Record the date and time exactly so that you can determine the day of scarifying.
6. This is typically done before 10 am to ensure the identification of all mated females.
7. Be careful not to damage the uterine horns.
8. Cells will exhibit bright colonies when they get optimal growth. If the color of medium becomes yellow, immediately change the medium to bring the pH back to neutral.
9. Avoid overgrowth because the ESCs will start to differentiate under overgrowth.
10. Allow 2–5 min until this happens. Avoid long time incubation with trypsin so the cells not to die.
11. The ESCs should appear as separate colonies and rounded.
12. At this point, ES cells are ready for any experiments scheduled or otherwise for passage at a 1:3 ration.

References

Axelrod HR. Embryonic stem cell lines derived from blastocysts by a simplified technique. Dev Biol. 1984;101:225–8.

Doetschman TC, Eistattaer H, Katz M, Schmidt W, Kemler R. The in vitro development of blastocyst derived embryonic stem cell lines: formation of yolk sac, blood islands and myocardium. J Embryol Exp Morphol. 1985;87:27–45.

Gossler A, Doetschman T, Korn R, Serfling E, Kemler R. Transgenesis by means of blastocyst derived embryonic stem cell lines. Proc Natl Acad Sci. 1986;USA83:9065–9.

Reubinoff BE, Pera MF, Fong C-Y, Trounson A, Bongso A. Embryonic stem cell lines from human blastocysts: somatic differentiation in vitro. Nat Biotechnol. 2000;18:399–404.

Shamblott MJ, Axelman J, Wang S, Bugg EM, Littlefield JW, Donovan PJ, et al. Derivation of pluripotent stem cells from cultured human primordial germ cells. Proc Natl Acad Sci. 1998; USA95:13,726–31.

Thomson JA, Itskovitz-Eldor J, Shapiro SS, Waknitz MA, Swiergiel JJ, Marshal VS, Jones JM. Embryonic stem cell lines derived from human blastocysts. Science. 1998;282:1145–7.

Thomson JA, Kalishman J, Golos TG, Durning M, Harris CP, Becker RA, Hearn JP. Isolation of a primate embryonic stem cell line. Proc Natl Acad Sci. 1995;USA92:7844–8.

Vans MJ, Kaufman MH. Establishment in culture of pluripotential cells from mouse embryos. Nature. 1981;292:154–6.

Wobus AM, Holzhausen H, Jakel P, Schneich J. Characterization of a pluripotent stem cell line derived from a mouse embryo. Exp Cell Res. 1984;52:212–9.

Chapter 4
Reprogramming of Normal Cells into Human Pluripotent Stem Cells

Abstract Human pluripotent stem cells (hPSCs) have the potential to become the source of materials for cell-based therapy. The similarity of hPSCs with the capability of human embryonic stem cells (hESCs) to self-renew while retaining their ability to differentiate into all cell types in the human body attracts scientists to think about replaceable options to avoid the ethical issue in using hESCs. According to this direction, reprograming normal cells into a pluripotent state has gathered enormous interest during the last 15 years since the Yamanaka group induced pluripotent stem cells from normal fibroblasts. In the process transduction of four transcription factors, Oct3/4, Sox2, Klf4, and c-Myc, were demonstrated effective on the reprograming. Later, many techniques of reprogramming normal cells have been established. One of them is the generation of human iPSCs from mononuclear cells (PBMCs) that are separated from peripheral blood by the infection of Sendai virus designed to express the four factors. The reprogramming process should be paid more attention to expect the expansion of the hPSCs availability of clinical and/or practical use. Here we will show the generation of hPSCs from peripheral blood separated from healthy donors. Separation of monocytes and reprogramming will be shown step by step through the process from taking blood samples to observation of hPSCs colonies on feeder cells.

Keywords Human pluripotent stem cells · Reprograming

4.1 Introduction

Various methods have been developed for reprogramming somatic cells into pluripotent cells, including nuclear transfer from somatic cells to oocytes, cell fusion with factors expressed in pluripotent cells, integrating factors into chromatin of somatic cells, and direct reprogramming (Patel and Yang 2010). Somatic cells have recently been shown to be reprogrammed by the introduction of selective transcription factors. These factors include Oct4, Sox2, c-myc, Klf4, Fbx15, and Nanog. By introducing these factors or other factors in combination with these via viral or non-viral vectors, somatic cells are directly reprogrammed into pluripotent cells. During the reprogramming, epigenetic events such as DNA methylation, histone

S. M. Afify, M. Seno, *Methods in Cancer Stem Cell Biology*,
https://doi.org/10.1007/978-981-99-1331-2_4

modification, and chromatin structure changes are required to exhibit phenotypes of embryonic stages. Although the recent studies appear to shed light on the roles of the transcription factors to promote pluripotency, the force to drive a somatic cell to pluripotent cells is not well explained or described by their roles of each transcription factor.

Oct4, Sox2, and Klf4 have been shown to work in combination to control expression and repression of a set of genes to maintain a pluripotent cell (Loh et al. 2006; Kim et al. 2008). The expression of these factors including c-myc may lead to a sequence of epigenetic events which influence the modification of chromatin and the methylation of DNA leading to the induction of pluripotent cells. Once these factors are introduced, the phenotype of a somatic cell will convert to a partially or totally reprogrammed state (Wernig et al. 2007; Okita et al. 2007).

Rendering the cells a property similar to pluripotent cells, c-myc proteins have been shown to be involved in loosening chromatin structure of somatic cells via histone acetylation (Dancy and Cole 2015). This structure allows Oct4 and Sox2 as well as Klf4 to bind to their target genes to initiate the expression of a key set of embryonic stem cell genes in somatic cells (Wernig et al. 2007). Oct4 and Sox2 then establish an autoregulatory loop maintaining the pluripotent state in the reprogrammed somatic cells (Masui et al. 2007).

One of the most commonly used transcription factor is Oct4 in the reprogramming because Oct4 plays an essential role as a major regulator of pluripotency (Boiani et al. 2002; Cheng et al. 2007).

Human induced pluripotent stem (iPS) cells resulting from the reprogramming of somatic cells are expected to be applied to regenerate organs of patients with some organ-specific disease. On the other hand, iPS cells could also be available to understand mechanisms of disease, to screen effective and safe drugs, and to ultimately treat various diseases and injuries. Since human embryonic stem cells are ethically difficult to obtain, reprogrammed somatic cells could perfectly be an effective alternative. In the current chapter we are trying to show the reprogramming process step by step from peripheral blood monocytes.

4.2 Materials

4.2.1 Reagents

- ROCK inhibitor (Y-27632, Sigma Aldrich, cat. no. SCM075).
- Mitomycin C treated mouse embryonic fibroblast (MEF) cells (REPROCELL Inc., Kanagawa, Japan).
- Basic FGF (Chemicon, CA, USA).
- SeVdp-iPS vector (SeVdp(KOSM) or SeVdp(KOSM)302L) 10^6–3×10^7 ciu (cell infectious unit)/mL in DMEM plus 10% FCS.
- siRNA L527 (10 μM mixture in RNase-Free water) L527 siRNA #1:

5′-GGUUCAGCAUCAAAUAUGAAG-3′ L527 siRNA #2: 5′- -UCAUAUUUGAUGCUGAACCAU-3′.
- TrypLE™ Express Enzyme (1×), no phenol red (Gibco™12604013).
- ROCK inhibitor, Y27632 (10 mM).
- Trypsin-EDTA (0.25%) (Nacalai Tesque, Kyoto, Japan, Cat. No: 327777-44).
- Penicillin/streptomycin mixed solution (100 U/mL) Nacalai Tesque, Kyoto, Japan, Cat. No-26253-84).
- 70% ethanol (Sigma-Aldrich; Cat. No.: 459836-2).
- TrypLE™ Express Enzyme (1×), no phenol red (Gibco™12604013).
- Dulbecco's Modified Eagle Medium (DMEM) (Thermo Fischer Scientific, MA).
- Minimum Essential Media (MEM) non-essential amino acid solution 100× (Wako-Fujifilm, Japan).
- Fetal bovine serum (FBS) (Gibco Life Technologies, MA).
- Penicillin/streptomycin mixed solution (Nacalai Tesque, Japan).
- L-glutamine (Nacalai Tesque, Japan).
- 2-Mercaptoethanol (Sigma-Aldrich, MO).
- Trypsin-EDTA solution. 0.25% trypsin (Sigma-Aldrich, MO).
- Phosphate buffered saline (PBS) (Genesee Scientific, CA).
- Hank's balanced salt solution (HBSS) (Genesee Scientific, CA).
- Ethanol (Sigma-Aldrich, MO).
- Blocking buffer: HBSS containing 1% BSA.
- FACS buffer: in HBSS containing 0.2% BSA.
- $CaCl_2$ (Sigma-Aldrich, MO).
- Dimethyl sulfoxide (DMSO) (Sigma-Aldrich, MO).
- Appropriate tissue culture plates and supplies.
- Povidone Iodine Scrub Solution 7.5% Iodine (Dynarex, NY).
- Isoflurane (WAKO-Fujifilm, Japan).
- BALB/c-nu/nu immunodeficient mice, 4-week old female (Charles River laboratories, Japan).

4.2.2 *Equipment*

- 37 °C water bath.
- Cell counting instrument (e.g., Vi-Cell™ or hemocytometer).
- Sterile cotton buds.
- CO_2 incubator MCO-19AIC(UV) (Panasonic, Japan).
- Bio-Clean Bench CCV-1307E (Hitachi, Japan).
- Olympus IX81 inverted microscope (Olympus, Japan).
- Laser scanning confocal microscope FV-1000 (Olympus, Japan).
- Tissue culture-treated plate, 60-mm dish (TPP Techno Plastic Products AG, Switzerland).
- Tissue culture-treated plate, 100-mm dish (TPP Techno Plastic Products AG, Switzerland).

- Falcon® Conical 15-mL centrifuge tubes (BD Falcon, NY).
- Falcon® Conical 50-mL centrifuge tubes (BD Falcon, NY).
- Cell Strainer 70 μm (BD Falcon, NY)
- Liquid N_2 storage tank.
- Microscope DIAPHOT 200 (Nikon, Japan).
- 1.5-mL microcentrifuge tubes (Eppendorf, Germany).
- Bench-top Hi-speed centrifuge 5810 R (Eppendorf, Germany).
- Eppendorf Safe-Lock microcentrifuge 1.5-mL tubes (Eppendorf, Germany).
- Sterile plastic disposable pipettes, 5/10/25-mL (Falcon, NY).

4.2.3 Reagent Preparation

Rock Inhibitor Solution

- Add sterile cell culture to make 1 mM stock solution.
- Dividing into small volume aliquots is recommended (100-μL aliquots/1.5 Eppendorf tube).
- Use working ROCKi concentrations of 10 μM.

Repro Stem Cell Medium

• Mix the following for 500 mL	
• Repro stem cell medium (#RCHEMD001)	495 mL
• BFGF, 0.5 μg/mL (final 5 ng/mL)	5 mL

StemFit Medium

• Mix the following for 500 mL	
• StemFitAK02NBasal (RCAK02N)	495 mL
• BFGF, 0.5 μg/mL (final 5 ng/mL)	5 mL

siRNA Complex Preparation

• Mix the following	
• 10 μM L527 siRNA (10 μL/a 6-well)	
• RNAiMax	(5μL/a 6-well)
• Opti-MEM	(125μL/a 6-well)
• Keep for 15 min at 25 °C	

MEF Medium (DMEM + 10% FCS)

• Mix the following	
• DMEM (Sigma: #D5796)	445 mL

(continued)

• FBS	50 mL
• Penicillin/streptomycin (Life Technologies)	5 mL

human embryonic stem cell (hES) Medium (1)

• Mix the following for 500 mL	
• Primate ES cell culture medium (ReproCell: #RCHEMD001)	495 mL
• FGF, 0.5 mg/mL (final 5 ng/mL)	5 mL

hES Medium (2)

• Mix the following for 500 mL	
• DMEM/F12 (Sigma: #D8437)	385 mL
• KnockOut Serum Replacement (Life Technologies)	100 mL
• Non-essential amino acids (Life Technologies)	5 mL
• 2-Mercaptoethanol (Life Technologies) (final 0.1 mM)	910 μL
• bFGF, 0.5 μg/ml (final 5 ng/ml)	5 mL
• Penicillin/streptomycin (Life Technologies)	5 mL

Dissociation Solution

• Mix the following for 100 mL	
• Trypsin (Difco; #215240)	0.25 g
• Collagenase IV, 10 mg/mL (Life Technologies, #17104-019)	10 mL
• KnockOut Serum Replacement (Life Technologies)	20 mL
• 1 M CaCl2 (final 1 mM)	0.1 mL
• PBS	70 mL

4.3 Methods

Our study focuses on monocyte separation and reprogramming into human iPSCs using the SeVdp-iPS vector SeVdp (KOSM) (Fig. 4.1). The use of SeVdp-iPS vector SeVdp (KOSM) in the generation of human iPS cells is deemed to be one of the most important methods for the generation of iPS cells. The reprogrammed human iPS cells can self-renew and differentiate into three germ layers in vivo.

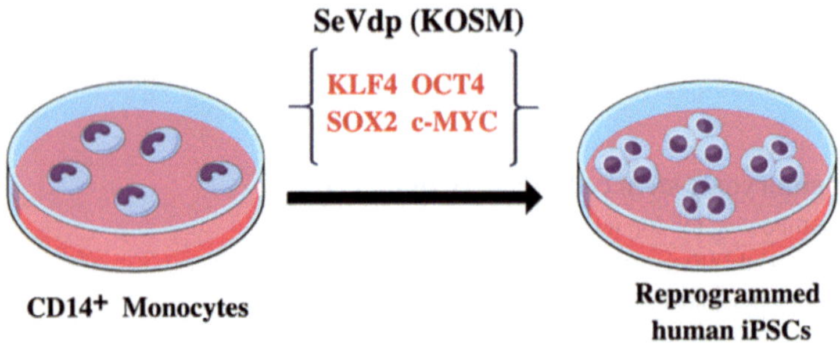

Fig. 4.1 Scheme for direct reprogramming human iPS cell from monocytes using SeVdp

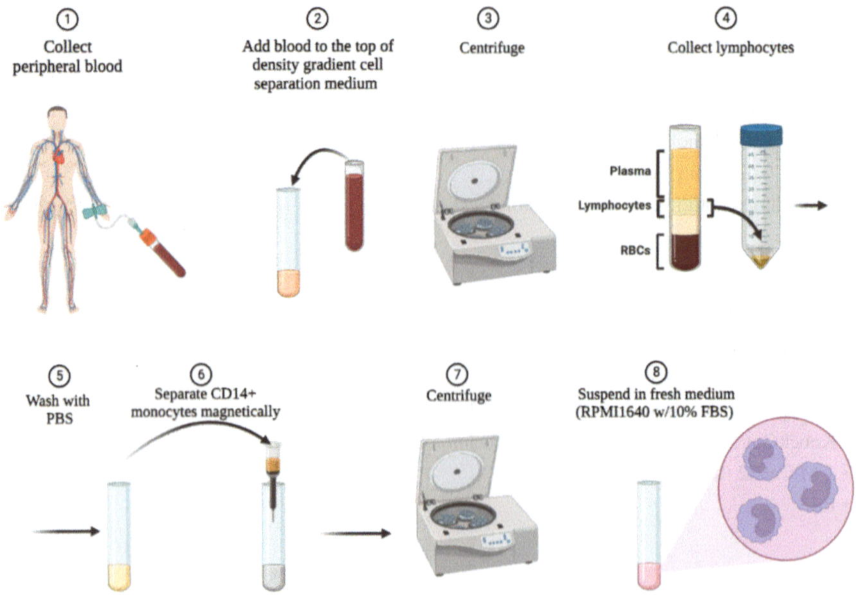

Fig. 4.2 Illustration of the Ficoll density gradient method for the separation of mononuclear cells from whole blood

4.3.1 Preparation of Healthy Donor and Patient peripheral blood mononuclear cells (PBMC) Samples

4.3.1.1 Ficoll Density Gradient Separation Method

The separation of whole blood by Ficoll density gradient is a generally common method to separate mononuclear cells (Fig. 4.2).

1. Collect fresh blood with heparin in 50-mL conical centrifuge tubes.
2. Put the tubes in a laminar flow bench.

3. Transfer blood from a collection tube into a 50-mL tube.
4. Add equal volume of PBS (1:1 dilution) using a sterile pipette.
5. Mix well by pipetting.
6. Prepare a new 50-mL tube with 15 mL of Ficoll medium.
7. Layer 15 mL of diluted whole blood over a Ficoll medium (Note 1).
8. Centrifuge the tube of blood on Ficoll medium bed for 30 min at 1500 rpm at 25 °C.
9. Remove the tubes carefully from the centrifuge without disturbing the layering.
10. Remove the PBMC layer from the tube and transfer to a new sterile 50-mL conical tube.
11. Increase the volume up to 15 mL with PBS.
12. Centrifuge at 1500 rpm for 10 min at 25 °C.
13. Discard the supernatant.
14. Loosen the pelleted cells.
15. Wash the pellet once again with PBS.
16. Discard the supernatant.
17. Resuspend the cells in an appropriate volume of PBS for subsequent assay.

4.3.1.2 Cell Preparation Tubes

1. Keep the BD Vacutainer® CPT™ Tube with Sodium Citrate at 25 °C (Note: Do not forget proper labels of individual names).
2. Collect fresh blood with heparin in 50-mL conical centrifuge tubes (Note: Stand the tubes in upright position until centrifugation at 25 °C).
3. Centrifuge the tubes of blood samples at 25 °C in a swing rotor for more than 20 min at 1500–1800 RCF (Note 2).
4. Open the tube in a laminar flow clean bench.
5. Aspirate approximately half of the plasma without disturbing the cell layer (Note: Mononuclear cells and platelets are in a white layer just under the plasma layer) (Fig. 4.2).
6. Collect the cell layer with a Pasteur pipette.
7. Transfer the cells to a sterile 50-mL conical tube.
8. Add PBS up to 15 mL.
9. Mix cells well.
10. Centrifuge at 400 × g (1500 rpm) for 10 min.
11. Aspirate supernatant as much as possible without disturbing cell pellet.
12. Resuspend cell pellet by gentle vortex mixing or tapping tube with a finger.
13. Add PBS up to 15 mL and repeat Steps 9–12.
14. Add an appropriate volume of PBS for subsequent assay or procedure.

4.3.2 Preparation of Gelatin-Coated Dishes

1. Transfer the 0.2% gelatin solution from 4 °C to the clean bench.
2. Transfer enough volume of 0.2% gelatin solution to cover the bottom of the dish (Note 3).
3. Incubate the dishes for 30 min at 37 °C in a 5% CO_2 incubator.

4.3.3 Preparation of MEF-Coated Dishes

1. Pick up mitotically inactivated MEF vial from liquid nitrogen storage using metal forceps and put on ice.
2. Wipe the vial with tissue paper.
3. Immerse the vial quickly in a 37 °C water bath without submerging the cap.
4. Remove the vial from the water bath just before the ice completely dissolves.
5. Detox the surface of the vial with 70% ethanol.
6. Take the vial into the laminar hood.
7. Add 5 mL of pre-warmed MEF medium to the cells in the 15-mL conical tube.
8. Pipette up the thawed cells gently and transfer into a sterile 15-mL conical tube containing MEF medium (Note 4).
9. Mix the cells gently by slow pipetting in the MEF medium.
10. Centrifuge the cells at 200 × g for 5 min.
11. Aspirate the supernatant.
12. Resuspend the cell pellet in MEF medium.
13. Count the cells using hemocytometer.
14. Pick up the gelatin-coated 6-well plate from the incubator.
15. Aspirate the excess gelatin solution.
16. Seed 1×10^6 cells of MEF/well.
17. Add 3 mL of MEF medium into each well.
18. Incubate the plates in a 37 °C in a 5% CO_2 incubator.
19. Use the plates with MEF feeder layer (Fig. 4.3) within 3–4 days after seeding (Note 5).

4.3.4 Generation of iPS Cells

Day 0

1. Prepare the cells at 3×10^5 cells/well in a 6-well plate (Note 6).
2. Magnetically separate $CD14^+$ monocytes.
3. Suspend 1×10^6 cells in 300 μL of fresh medium (RPMI1640 w/10% FBS) containing SeVdp-iPS vector SeVdp(KOSM)302L (10^6–3×10^7 cell infectious unit/mL) at MOI 2.

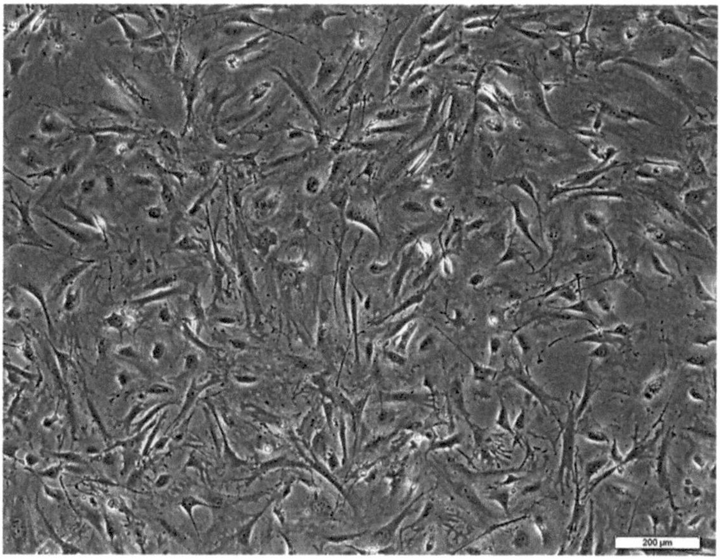

Fig. 4.3 Representative image of MEF feeder cells on gelatin-coated well

4. Incubate at 25 °C (or at 37 °C, 5% CO2) for 2 h.
5. Wash the cells once with RPMI1640 w/10% FBS.
6. Centrifuge for 5 min at 200 × g.
7. Suspend the cells in RPMI1640 w/10% FBS.
8. Seed the cells on the MEF feeder cells at 1×10^6 cells/well in a 6-well plate.
9. Incubate the cells at 37 °C 5% CO_2 incubator in RPMI1640 containing 10% FBS.

Day 3

1. Change medium by pre-warming human ES medium (Note: human ES medium is primate ES cell culture medium (#RCHEMD001, ReproCell, Japan) containing 5 ng/mL bFGF)
2. Change the medium every 2 days (Note 7).

Day 8 : On day 8, we can choose one method from the following: (A) Bulk Method

1. Prepare three 6-well plates coated with MEF.
2. Aspirate the supernatant from the colonies of human iPSCs.
3. Wash the colony with 1× PBS.
4. Dissociate the cells with TrypLE (Note 8).
5. Seed each dish in MEF-coated 6 well plates.
6. Add ROCK inhibitor (10 μM Y27632) to enable maintenance of stem cell phenotype and prevent the cell death.
7. Add 515 μL of siRNA complex (10 μM of each 5′-GGUUCAGCAU CAAAUAUGAAG-3′ and 5′-UCAUAUUUGAUGCUGAACCAU-3′)/2 mL medium to remove SeVdp-iPS genome (Note 9).

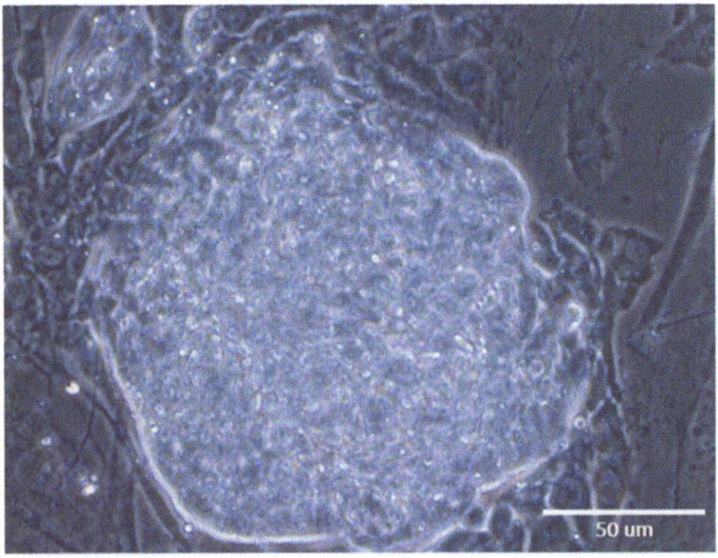

Fig. 4.4 Representative image of human iPSCs on MEF. Scale bars = 50 μm

8. Every 2 days, replace the medium with a fresh hES medium containing siRNA complex. (Note 10).

(B) Colony Isolation Method

1. Wash the cells once with PBS.
2. Add fresh human ES medium.
3. Cut the periphery of the colony with a 24-gage needle.
4. Pick-up the colony by micropipette.
5. Transfer the colony to a 1.5 mL-tube containing 50 μL of human ES medium.
6. Add 500 μL of pre-warmed human ES medium to the tube.
7. Dissociate the cells mechanically by passing through a 26-gage needle.
8. Add ROCK inhibitors to prevent cell death.
9. Add 128 μL of siRNA complex/well.
10. Change medium on the next day with 500 μL of fresh medium containing siRNA every 2 days (Note 11).
11. Change medium every 2 days and monitor human iPS colonies (Fig. 4.4).
12. After 3 weeks hiPS free SeVdp will develop and could be stained against stemness markers such as NANOG, SOX2, and OCT-4 (Fig. 4.5).

Notes

1. Hold the tube with an angle and pipette the blood gently along with the wall of the tube so as not to disturb the top of the separation medium.
2. Samples should be centrifuged within 2 h after collection.

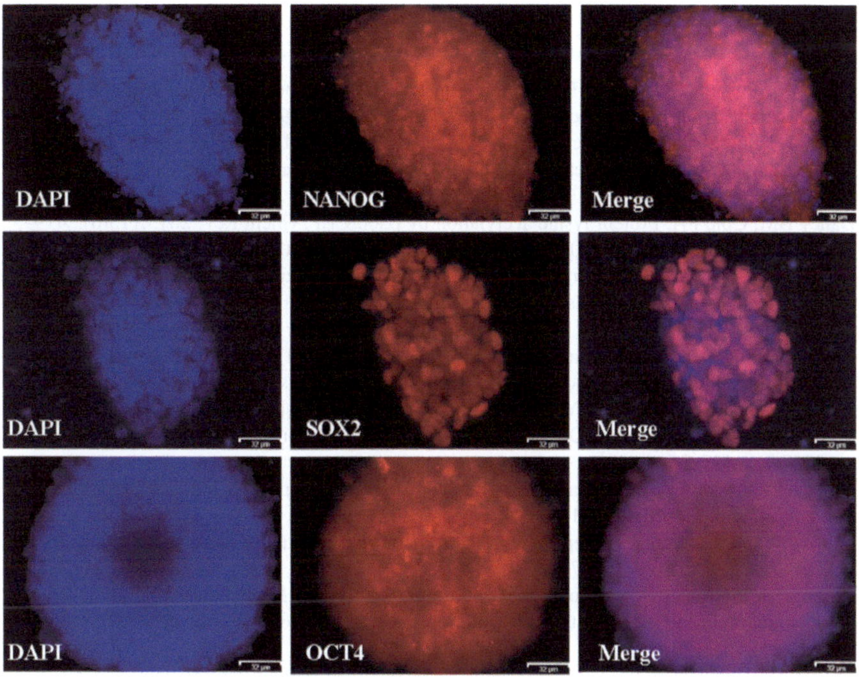

Fig. 4.5 Representative immunofluorescent images of human iPS stained with the antibodies against Nanog, SOX2, and OCT4. Scale bars = 32 μm

3. 2 mL for 60-mm dishes and 5 mL for 100-mm dishes.
4. This step should be done as quickly as possible so the cells do not get stressed.
5. Plates with MEF feeder layer can be kept up to approx. 2 weeks in the incubator until use.
6. In the case of blood, the cells are PBMC, CD14+ monocytes, T-cells, CD34+ cord blood cells, CD34+ peripheral blood cells, etc.
7. If there are colonies detached during the medium, you should centrifuge and plate again. Generally, in 6 days, colonies of human iPSCs will appear.
8. Keep mentoring the colonies, do not wait until the colonies dissociate in single cells.
9. Removal of SeVdp-iPS genome is essential for the generation of transgene-free iPS cells. Removal of SeVdp (KOSM) genome requires siRNA treatment. Removal of SeVdp (KOSM)302L genome is semi-automatic in response to induction of mir302, but siRNA treatment is still recommended to facilitate the removal.
10. Repeat this step 4 times.
11. Repeat this step 3–4 times.

References

Boiani M, Eckardt S, Scholer HR, McLaughlin KJ. Oct4 distribution and level in mouse clones: consequences for pluripotency. Genes Dev. 2002;16(10):1209–19.

Cheng L, Sung MT, Cossu-Rocca P, Jones TD, MacLennan GT, De Jong J, et al. OCT4: biological functions and clinical applications as a marker of germ cell neoplasia. J Pathol. 2007;211(1): 1–9.

Dancy BM, Cole PA. Protein lysine acetylation by p300/CBP. Chem Rev. 2015;115(6):2419–52. https://doi.org/10.1021/cr500452k. Erratum in: Chem Rev. 2016 Jul 27;116(14):8314

Kim JB, Zaehres H, Wu G, Gentile L, Ko K, Sebastiano V, et al. Pluripotent stem cells induced from adult neural stem cells by reprogramming with two factors. Nature. 2008;454(7204): 646–50.

Loh YH, Wu Q, Chew JL, Vega VB, Zhang W, Chen X, et al. The Oct4 and Nanog transcription network regulates pluripotency in mouse embryonic stem cells. Nat Genet. 2006;38(4):431–40.

Masui S, Nakatake Y, Toyooka Y, Shimosato D, Yagi R, Takahashi K, et al. Pluripotency governed by Sox2 via regulation of Oct3/4 expression in mouse embryonic stem cells. Nat Cell Biol. 2007;9(6):625–35.

Okita K, Ichisaka T, Yamanaka S. Generation of germline-competent induced pluripotent stem cells. Nature. 2007;448(7151):313–7.

Patel M, Yang S. Advances in reprogramming somatic cells to induced pluripotent stem cells. Stem Cell Rev Rep. 2010;6(3):367–80.

Wernig M, Meissner A, Foreman R, Brambrink T, Ku M, Hochedlinger K, et al. In vitro reprogramming of fibroblasts into a pluripotent ES-cell-like state. Nature. 2007;448(7151): 318–24.

Chapter 5
Maintenance of Human Pluripotent Stem Cells

Abstract During the last decades, the field of stem cell research has rapidly expanded with the potential to develop the technology of tissue regeneration as well as the discovery of novel therapeutic agents. The innovative procedure of cellular reprogramming, which was found by Yamanaka in 2007, lead to the generation of induced pluripotent stem cells (iPSCs) from a somatic cell namely a normal fibroblast. The maintenance of iPSCs in a successful manner is critical issue because undifferentiated state should be kept in the culture. Especially, extra attention should be payed to maintain their key characteristics of self-renewal and pluripotency avoiding unexpected differentiation. In this chapter, we describe the basic techniques necessary to culture human or mouse iPSCs, e.g starting from the frozen stocks, eeding cells into culture vessels, changing media, passaging, and cryopreservation.

Keywords Pluripotency · Stem cells · Maintenance

5.1 Introduction

Cells in the embryonic stem cell (ESC) mass including those in the embryonic inner cell mass, are pluripotent. It is widely accepted that pluripotent cells are capable to generate all somatic cells (Solter 2006; Bioani and Schöler 2006). A number of studies have revealed the pluripotency depends on epigenetic processes and transcription factor networks (Niwa et al. 2000; Boyer et al. 2006). A combination of differentiation prevention and proliferation is necessary to maintain the pluripotency during cell division. A Prevention of the differentiation allows ESCs to continuously self-renew for a long period

The culture conditions used combinations of different small molecules and growth factors to control the extrinsic signaling pathways which are known to play critical roles in the differentiation of stem cells. One of the most important growth factors for keeping mouse ES cells undifferentiated and proliferating is leukemia inhibitory factor (LIF). In terms of preventing differentiation, LIF plays a crucial role. In the interleukin-6 family, LIF binds to the heterodimeric receptor composed of, LIF receptor, and gp130 (also known as IL6ST, IL6R-beta or CD130).

S. M. Afify, M. Seno, *Methods in Cancer Stem Cell Biology*,
https://doi.org/10.1007/978-981-99-1331-2_5

On the binding of LIF to the receptor, the Jak/Stat pathway is activated. The activation of Stat3 has been demonstrated essential and sufficient to maintain pluripotency in ESCs (Niwa et al. 1998; Matsuda et al. 1999). The mechanisms that prevent differentiation of ESCs and promote their proliferation must be taken over to their daughter cells as far as LIF stimulates them

The pluripotency in human iPSCs is orchestrated by fibroblast growth factors (FGFs). In order to maintain the pluripotency, the expression of the same subsets of the gene expressed in pluripotent blastula cells is controlled in iPSCs. The FGF family contains 22 members of polypeptides that are biochemically important in both structural and functional of the cells. There is an overall 30–50% homology in amino acid sequences between the members. FGFs have generally been characterized by an extraordinary high affinity for heparin/heparan sulfate and two conserved cysteine residues. Acidic FGF (aFGF/FGF1) and basic FGFs (bFGF/FGF2) are the first two members of FGF family identified by the different isoelectric points.

Several members of the FGF family including FGF2, FGF4, FGF6, FGF7, FGF8, and FGF9 have been reported to impact on the stemness. FGF2 and FGF4 are considered essential for maintaining mouse and human stem cells. The expression of all FGF receptors (FGFRs) in human embryonic stem cells (ESCs) has been previously reported. Interestingly, FGF2 stimulates all FGFRs as a pleiotropic ligand (Ding et al. 2010). A recent study has shown that FGF2 binding to FGFR1, the most important receptor, activates downstream signaling pathways of PI3K/AKT as well as small ras/raf/MAPK/ERK (Nakashima and Omasa 2016). A description of the basic techniques for culturing iPSCs is provided here, starting with frozen stocks and plating them into culture vessels, medium changes, passage, and preparation of frozen to stocks.

5.2 Materials

5.2.1 Reagents

- ROCK inhibitor (Y-27632, Sigma Aldrich, cat. #SCM075).
- Mitomycin C treated mouse embryonic fibroblast (MEF) cells (REPROCELL Inc., Kanagawa, Japan).
- Basic FGF (Chemicon, CA, USA).
- Trypsin-EDTA (0.25%) ((2.5g/l-Trypsin/1mmol/l-EDTA Solution, Nacalai Tesque, Kyoto, Japan, Cat. No: 327777-44).
- Penicillin (10,000 unit/ml)/streptomycin (10,000 μg/ml) mixed solution (100mL) Nacalai Tesque, Kyoto, Japan, cat #26253-84).
- 99.5% ethanol (FUJIFILM Wako Chemicals).
- TrypLE™ Express Enzyme (1×), no phenol red (Gibco™, Thermo Fisher, cat. #12604013).
- Dulbecco's Modified Eagle Medium (DMEM) high glucose (FUJIFILM Wako Chemicals).

- Minimum essential media (MEM) non-essential amino acid solution 100× (FUJIFILM Wako Chemicals).
- Fetal bovine serum (FBS) (Gibco Life Technologies, MA).
- L-glutamine (Nacalai Tesque, Japan).
- 2-Mercaptoethanol (Sigma-Aldrich, MO).
- Phosphate buffered saline (PBS) without CaCl2 or MgCl2 (Genesee Scientific, CA).
- Hank's balanced salt solution (HBSS) (Genesee Scientific, CA).
- Blocking buffer: HBSS containing 1% bovine serum albumin (BSA).
- FACS buffer: HBSS containing 0.2% BSA.
- $CaCl_2$ dihydrate (Sigma-Aldrich, MO).
- Dimethyl sulfoxide (DMSO) (Sigma-Aldrich, MO).
- Appropriate tissue culture plates and supplies.

5.2.2 Equipment

- 37 °C water bath.
- Cell counting instrument (e.g., hemocytometer, TC20 Automated Cell Counter (Bio-Rad) or Vi-Cell™ (Beckman).
- Standard forceps, 12–13 cm, straight, pointed.
- CO_2 incubator MCO-19AIC(UV) (Panasonic, Japan).
- Bio-Clean Bench CCV-1307E (Hitachi, Japan).
- Olympus IX81 inverted microscope (Olympus, Japan).
- Laser scanning confocal microscope FV-1000 (Olympus, Japan).
- Tissue culture dishes for adherent cells, 60-mm dish (TPP Techno Plastic Products AG, Switzerland).
- Tissue culture dishes for adherent cells, 100-mm dish (TPP Techno Plastic Products AG, Switzerland).
- ®Conical 15-mL centrifuge tubes (BD Falcon, NY).
- ®Conical 50-mL centrifuge tubes (BD Falcon, NY).
- Cell strainer 70 μm (BD Falcon, NY).
- Liquid N_2 storage tank.
- Inverted microscope CKX53 (NikonOlympus, Japan) with digital camera.
- 1.5-mL microcentrifuge tubes (Eppendorf, Germany).
- Bench-top high-speed centrifuge 5810 R (Eppendorf, Germany).
- Eppendorf safe-lock microcentrifuge 1.5-mL tubes (Eppendorf, Germany).
- Sterile plastic disposable pipettes, 5/10/25-mL (Falcon, NY).

5.2.3 *Reagent Preparation*

Rock Inhibitor Solution

- Add sterile PBS to make 10 mM stock solution.
- Dividing into small volume aliquots is recommended (100-μL aliquots/1.5-ml Eppendorf tube).
- Use ROCKi at 10 μM as a working concentration.

Repro Stem Cell Medium

• Mix the following per 500 mL	
• Repro stem cell medium (#RCHEMD001)	495 mL
• basic FGF, 0.5 μg/mL (final 5 ng/mL)	5 mL

StemFit Medium

• Mix the following per 500 mL	
• StemFit AK02N Basal (RCAK02N)	495 mL
• basicFGF, 0.5 μg/mL (final 5 ng/mL)	5 mL

Dissociation Solution

• Mix the following per 100 mL	
• Trypsin (Difco; #215240)	0.25 g
• Collagenase IV, 10 mg/mL (Life Technologies, #17104-019)	10 mL
• KnockOut™ Serum Replacement (Life Technologies)	20 mL
• 1 M CaCl2 (final 1 mM)	0.1 mL
• PBS	70 mL

5.3 Methods

In Vitro Culture Conditions for Human iPSCs

The optimal growth conditions make it possible to produce high-quality iPSCs (Takahashi and Yamanaka 2006). A similar time frame to ESCs is found in iPSCs, 18–20 h for cell doubling. However, if the density of the iPSCs is too high, they tend to grow as 3D cellular aggregates (Fig. 5.1). It is generally common for cellular aggregates of iPSCs to provide cells with heterogeneous subpopulation, which may be difficult to distinguish from parental iPSCs. To maintain the growth of initial iPSCs, cells should be passaged every 4–5 days. Periodical cloning of iPSCs is also recommended if the growth becomes slow.

5.3.1 Preparation of Gelatin-Coated Dishes

1. Transfer 0.2% gelatin solution from 4 °C to the clean bench.
2. Cover the bottom of the dish with sufficient volume of 0.2% gelatin solution (Note 1).
3. Incubate the dishes for 30 min at 37 °C in a 5% CO_2 incubator.

5.3.2 Preparation of MEF-Coated Dishes

1. Pick up a storage vial of mitotically inactivated MEF from liquid nitrogen.
2. Immerse the vial quickly in a 37 °C water bath without submerging the cap.
3. Remove the vial from the water bath just before the ice completely dissolves.
4. Sterilize the surface of the vial with 70% ethanol.
5. Put the vial into the laminar hood.
6. Prepare 5 mL of pre-warmed MEF medium in a sterile 15-mL conical tube.
7. Pipette up the thawed cells gently and transfer into the conical tube containing MEF medium (Note 2).
8. Mix the cells by slow and gentle pipetting in the MEF medium.

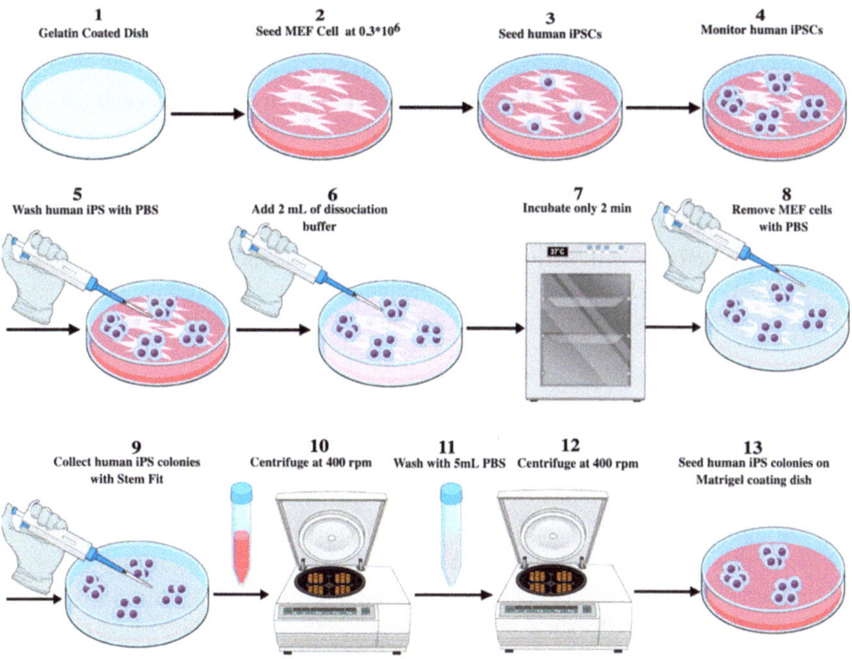

Fig. 5.1 An illustration of maintaining human iPSCs on MEF cells and how to transfer them to a feeder-less dish coated with Matrigel

9. Centrifuge the cells at 200 × g for 5 min at 25 °C.
10. Aspirate the supernatant.
11. Resuspend the cell pellet with 2 mL of fresh MEF medium.
12. Count the cells using hemocytometer.
13. Pick up the gelatin-coated 6-well plate from the incubator.
14. Aspirate the gelatin solution.
15. Seed MEF at 1×10^6 cells/well.
16. Add 3 mL of MEF medium per well.
17. Incubate the plates at 37 °C in a 5% CO_2 incubator.
18. Use the plates with MEF feeder layer within 3–4 days after seeding (Note 3).

5.3.3 Reviving Human iPSCs

1. Pre-warm the stem cell medium at 37 °C in a water bath for 30 min (Note 4).
2. Pick up the vial of human iPSCs from liquid nitrogen storage.
3. Thaw the cells quickly at 37 °C in a water bath.
4. Sterilize the vial with 70% ethanol.
5. Add 5 mL of pre-warmed stem cell medium to a sterile 15-mL conical tube.
6. Suspend the cells gently with a sterile long Pasteur pipette.
7. Transfer the cells into the 15-mL conical tube containing the medium.
8. Centrifuge cells at 200 × g for 5 min at 25 °C.
9. Aspirate the supernatant without disturbing the pellet.
10. Add 5 mL of pre-warmed stem cell medium.
11. Resuspend the pellet by gentle pipetting up and down 2 or 3 times using a sterile long Pasteur pipette.
12. Pick up the MEF-coated dishes prepared in the previous step from the incubator.
13. Replace the MEF medium with 4 mL of Repro stem medium containing 5 ng/mL of human FGF2 and 10 μM of rock inhibitor Y-27632.
14. Seed the cells on two MEF-coated dishes with high and low numbers (Note 5).
15. Remove the dead cells by changing the medium on the next day.
16. Monitor the cells until the colonies become visible and bright (Fig. 5.2).

5.3.4 Passage of Human iPSCs on MEF

1. Passage human iPSCs when the colonies become large and bright (Note 6).
2. Take the dish out from the incubator.
3. Aspirate the medium.
4. Wash the dish twice with 2 mL of PBS.
5. Aspirate PBS.
6. Add 250 μL of dissociation buffer/per a 60-mm dish.
7. Incubate the dish for 5 min in the incubator at 37 °C under 5% CO_2.

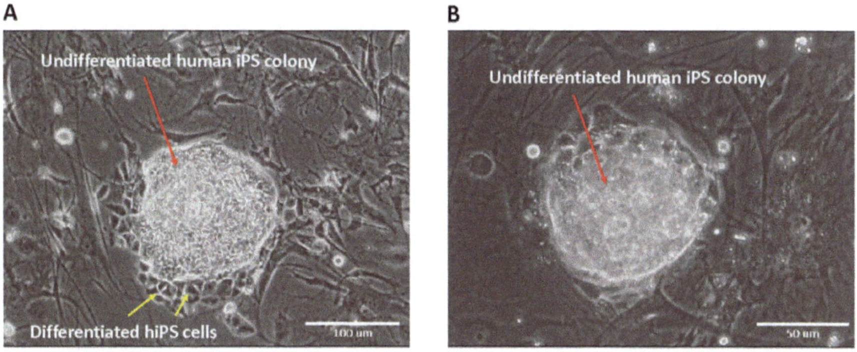

Fig. 5.2 Typical colonies of human iPSCs growing on MEF cells. The colony mainly shapes round on the edges with little (a) or no differentiation (b). Avoid differentiation of human iPSCs during the passages to obtain necessary number of cells.

8. Take the dishes out from the incubator.
9. Aspirate the dissociation buffer.
10. Wash the colonies of human iPSCs with PBS.
11. Add 1 mL of pre-warmed stem culture medium.
12. Detach the colonies of human iPSCs with a cell scraper.
13. Collect the colonies and make them small by gentle pipetting.
14. Take MEF-coated dishes out from the incubator.
15. Aspirate MEF medium.
16. Add 4 mL Repro Stem medium.
17. Divide the small colonies into MEF-coated dishes.
18. Add FGF2 to 5 ng/mL and rock inhibitor, Y-27632, to 10 μM into every 5-mL Repro Stem medium.
19. Incubate the dishes in the incubator at 37 °C under 5% CO_2.
20. Change the medium to fresh Repro Stem medium without Y-27632 on the next day (Fig. 5.3).

5.3.5 Passage of Human iPSCs on Matrigel

1. Pick up Matrigel from a freezer and place at 4 °C 1 day before transferring human iPSCs.
2. Add 2 mL of Repro Stem medium into a 15-mL tube placed on ice.
3. Add 50 μL of Matrigel with cold tips.
4. Add 2 mL of Repro Stem medium into a 60-mm dish containing Matrigel.
5. Incubate the dish in an incubator at 37 °C with 5% CO_2 for at least 1 h.
6. Take the 60-mm dish of human iPSCs out from the incubator.
7. Aspirate the medium from the dish.

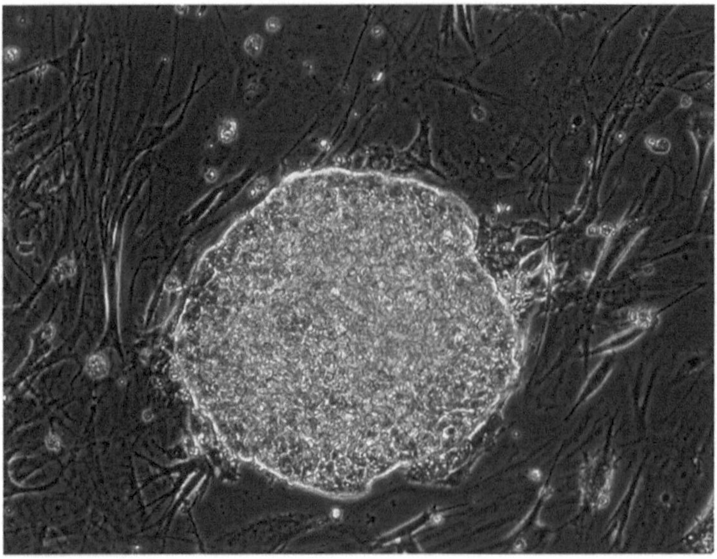

Fig. 5.3 Human iPSCs on MEF cells after the second passage selecting the undifferentiated colonies from the revived cells.

8. Wash the dish twice with PBS.
9. Add 250 μL of dissociation buffer to the dish.
10. Spread the buffer well over the surface.
11. Incubate the dish in the incubator at 37 °C with 5% CO_2 for 2–3 min until MEF cells detach.
12. Aspirate the dissociation buffer.
13. Wash the colonies of human iPSCs with PBS.
14. Detach the colonies of human iPSCs with a cell scraper.
15. Collect the colonies and make them small by gentle pipetting.
16. Take the Matrigel-coated dish out from the incubator.
17. Aspirate media from the dish.
18. Wash the dish with PBS.
19. Add 4 mL of StemFit media into the dish.
20. Divide human iPSC colonies into three Matrigel-coated dishes.
21. Add FGF2 to 5 ng/mL and rock inhibitor, Y-27632, to 10 μM into every 5-mL StemFit medium.
22. Incubate the dish in an incubator at 37 °C with 5% CO_2.
23. Change medium to StemFit medium without Rock inhibitor, Y-27632 on the next day (Fig. 5.4).

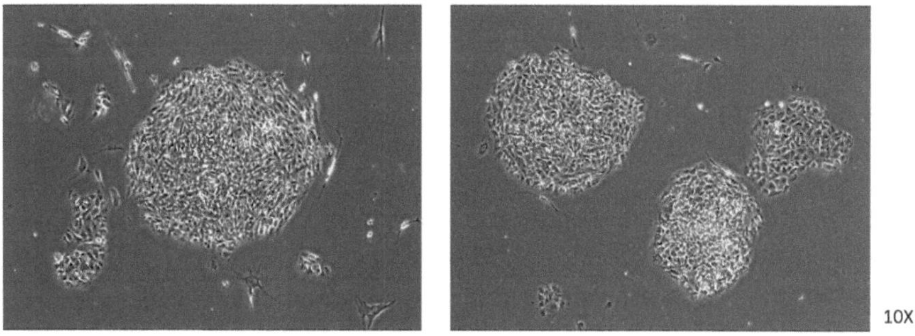

Fig. 5.4 Human iPSCs on Matrigel exhibiting typical hESC-like morphology, with round colonies with small diameters

5.3.6 Preparation of Cryopreservation Medium for iPSCs

Cryopreservation is one of the most important processes in the handling of human iPSCs. Cryopreservation medium: Dimethyl sulfoxide (DMSO) is an essential cryopreservant component of cryopreservation medium. The concentrations of DMSO should be as high as 10% to provide high cryopreservation efficiency in StemFit medium. Here are the steps you need to follow to freeze human iPSCs cells to make stocks.

1. Take the dish out from the incubator.
2. Aspirate the medium.
3. Wash the dish twice with 2 mL of PBS.
4. Aspirate PBS.
5. Add 250 μL of dissociation buffer/per a 60-mm dish.
6. Incubate the dish for 2 min in the incubator at 37 °C under 5% CO_2.
7. Take the dishes out from the incubator.
8. Collect human iPS colonies using scrapers (Note 7).
9. Add 5 mL of StemFit medium.
10. Transfer the cells into a sterile 15 mL-conical tube.
11. Centrifuge at 100 × g for 5 min.
12. Discard the supernatant without disturbing the pellet.
13. Add the cryopreservation medium containing 10% DMSO.
14. Transfer the cell suspension in to cryopreservation vial and put the vial into a freezing treatment container such as BICELL (Nihon Freezer Co., Ltd., Japan), then transfer to −80 °C.
15. After 24 h transfer the vails to liquid nitrogen.

Notes

1. For example, 2 mL for a 60-mm dish and 5 mL for a 100-mm dish.
2. This step should be done as quickly as possible so as not to get the cells stressed.

3. Plates with MEF feeder layer can be kept up to approximately 2 weeks in the incubator until use.
4. Do not keep the media in the water bath for more than 1 h at 37 °C as continued exposure to 37 °C may impair the components and reduce the activity of the growth factors.
5. Avoid seeding human iPSCs at a high density because they tend to aggregate and give rise to cells with heterogeneous morphologies.
6. Do not wait until the colonies become thick and dark because the cells will start to die.
7. Avoid dispersing single cells from human iPS colonies.

References

Bioani M, Schöler HR. Regulatory networks in embryo-derived pluripotent stem cells. Nat Rev Mol Cell Biol. 2006;6:872–84.

Boyer L, Plath K, Zeitlinger J, Brambrink T, Medeiros LA, Lee TI, Levine SS, Wernig M, Tajonar A, Ray MK, et al. Polycomb complexes repress developmental regulators in murine embryonic stem cells. Nature. 2006;441:349–53.

Ding VM, Ling L, Natarajan S, Yap MG, Cool SM, Choo AB. FGF-2 modulates Wnt signaling in undifferentiated hESC and iPS cells through activated PI3-K/GSK3beta signaling. J Cell Physiol. 2010;225:417–28.

Matsuda T, Nakamura T, Nakao K, Arai K, Katsuki M, Heike T, Yokota T. STAT3 activation is sufficient to maintain an undifferentiated state of mouse embryonic stem cells. EMBO J. 1999;18:4261–9.

Nakashima Y, Omasa T. What Kind of Signaling Maintains Pluripotency and Viability in Human-Induced Pluripotent Stem Cells Cultured on Laminin-511 with Serum-Free Medium? Biores Open Access. 2016;5(1):84-93.

Niwa H, Burdon T, Chambers I, Smith A. Self-renewal of pluripotent embryonic stem cells is mediated via activation of STAT3. Genes Dev. 1998;12:2048–60.

Niwa H, Miyazaki J, Smith AG. Quantitative expression of Oct-3/4 defines differentiation, dedifferentiation or self-renewal of ES cells. Nat Genet. 2000;24:372–6.

Solter D. From teratocarcinomas to embryonic stem cells and beyond: a history of embryonic stem cell research. Nat Rev Genet. 2006;7:319–27.

Takahashi K, Yamanaka S. Induction of pluripotent stem cells from mouse embryonic and adult fibroblast cultures by defined factors. Cell. 2006;126(4):663–76. https://doi.org/10.1016/j.cell.2006.07.024. Epub 2006 Aug 10. PMID: 16904174.

Chapter 6
Identification of Cancer Stem Cells by Different Molecular Markers

Abstract Cancer stem cells (CSCs), also known as tumor initiating cells (TICs), exist in all tumor cells as an undifferentiated subpopulation in the range from 0.1 to 10%. CSCs are considered quiescent while capable of continuous proliferation as self-renewal playing critical roles in tumor growth, metastasis, and recurrence. The expression level of the antigens specific to CSCs is relatively lower than that of the established tumor-associated antigens, which should depict the differentiated phenotypes, probably due to their stemness. On the other hand, CSCs could be identified by the cell surface markers specific to stem cells. As a result, the expression patterns of the markers are found variable depending on the tissues and/or organs. Combined with the molecular markers, CSCs have been identified by different characters. The ability to form tumor spheres is one of the most critical characters to identify CSCs. A new stage of cancer research as well as the development of novel therapeutic strategies will be led by understanding the significance of CSCs. In this chapter, we describe step by step the techniques to characterize the CSCs by evaluating the markers as the potential of self-renewal and differentiation.

Keywords Cancer stem cells · CSC markers · Molecular markers

6.1 Introduction

CSCs have been described in various cancers, including those originating in the breast, brain, blood (leukemia), skin, thyroid, lung, gastrointestinal tract, reproductive tract, and head and neck (Ayob and Ramasamy 2018). In general, the stemness of cells is judged by the presence or absence of expression of certain markers, many of which are also found expressed in CSCs (Karsten and Goletz 2013).

Some markers are available to distinguish embryonic stem cells from adult stem cells or pluripotent stem cells from progenitor cells. However, it is currently impossible to distinguish CSCs from normal stem cells by the expression pattern of stem cell markers. Based on the similarity of cell surface markers, CSCs could most likely originate from normal stem cells via epigenetic and genetic changes (Afify and Seno 2019; Afify et al. 2022). For example, onco-fetal stem cell markers, which are expressed in embryos/fetuses but not in adult tissues, and often expressed in cancer

S. M. Afify, M. Seno, *Methods in Cancer Stem Cell Biology*,
https://doi.org/10.1007/978-981-99-1331-2_6

cells, are currently explored as a preferred marker to identify CSCs (Karsten and Goletz 2013; Hsu et al. 2011).

Further, the number of surface markers available for the detection and characterization of CSCs has remarkably increased in recent years (Kim and Ryu 2017). Cell surface molecules are generally transmembrane proteins, which have pivotal roles in the cytoplasmic signal transduction and cell-to-cell communications. The expression profiles of cell surface markers vary depending on the cellular phenotypes on the stage of differentiation (Goodwin et al. 2020). Therefore, they are used to distinguish one cell type from the others and to sort cells and group them in narrow categories where each group could be explained with some common features. The cell surface markers have become important nowadays in the field of cell biology, especially in CSCs as well as stem cells.

CSCs have already been isolated from many types of cancers depending on the CSC-specific characteristics and surface markers. For instance, leukemic stem cells were the first type of CSCs, which were identified as $CD34^+/CD38^-$ cells together with the high tumorigenicity in vivo (Bonnet and Dick 1997). The isolation and enrichment of CSC have been achieved with a different panel of surface markers.

The well-accepted CSC surface markers include CD24, CD44, CD34, CD133, and EpCAM. The combination of CD24 and CD44 has been investigated as the breast CSC marker (Jaggupilli and Elkord 2012). Although various reports on the expression of the two markers appeared during the course of time, $CD44^+/CD24^-$ became fixed as the breast CSC marker. The expression levels of specific surface markers have been found higher in CSC populations than in non-CSC ones. It recently became apparent that the combination of different sets of these markers is more feasible for the characterization of CSCs than each of them alone. And the presence or absence of specific markers is critically essential to identify CSCs in different types of cancers (Walcher et al. 2020).

CSC surface markers are not only used to identify but are also available to isolate and/or enrich CSCs and finally to develop therapeutic strategies targeting CSCs.

Although some of these markers are still controversial and more investigation is needed to be used to judge CSCs, the number of CSC markers is considerably increasing in recent years. In the current chapter, we use fluorescence activated cell sorting (FCAS) to identify the CSC subpopulations in breast and liver cancer cell lines.

6.2 Materials

6.2.1 Reagents

- Human brain glioblastoma cell line U-251MG cell (ECACC, UK).
- Dulbecco's Modified Eagle Medium (DMEM) (Thermo Fisher Scientific, MA).
- Minimum essential media (MEM) non-essential amino acid solution 100× (Wako-Fujifilm, Japan).

- Fetal bovine serum (FBS) (Gibco Life Technologies, MA).
- Penicillin/streptomycin mixed solution (Nacalai Tesque, Japan).
- L-glutamine (Nacalai Tesque, Japan).
- 2-Mercaptoethanol (Sigma-Aldrich, MO).
- Trypsin-EDTA solution. 0.25% trypsin (Sigma-Aldrich, MO).
- Phosphate-buffered saline (PBS) (Genesee Scientific, CA).
- Hank's balanced salt solution (HBSS) (Genesee Scientific, CA).
- Ethanol (Sigma-Aldrich, MO).
- Blocking buffer: HBSS containing 1% BSA.
- FACS buffer: in HBSS containing 0.2% BSA.
- $CaCl_2$ (Sigma-Aldrich, MO).
- Dimethyl sulfoxide (DMSO) (Sigma-Aldrich, MO).
- Appropriate tissue culture plates and supplies.
- Povidone Iodine Scrub Solution 7.5% Iodine (Dynarex, NY).
- Isoflurane (WAKO-Fujifilm, Japan).
- BALB/c-nu/nu immunodeficient mice, 4-week-old female (Charles River laboratories, Japan).

6.2.2 *Equipment*

- Syringe needles, 24G.
- 37 °C water bath.
- Cell counting instrument (e.g., Vi-Cell™ or hemocytometer).
- Sterile cotton buds.
- Iris/eye scissors, straight.
- Operating scissors.
- Standard forceps, 12–13 cm, straight, pointed.
- CO_2 incubator MCO-19AIC (UV) (Panasonic, Japan).
- Bio-Clean Bench CCV-1307E (Hitachi, Japan).
- Olympus IX81 inverted microscope (Olympus, Japan).
- Laser scanning confocal microscope FV-1000 (Olympus, Japan).
- Tissue culture-treated plate, 60-mm dish (TPP Techno Plastic Products AG, Switzerland).
- Tissue culture-treated plate, 100-mm dish (TPP Techno Plastic Products AG, Switzerland).
- Falcon® Conical 15-mL centrifuge tubes (BD Falcon, NY).
- Falcon® Conical 50-mL centrifuge tubes (BD Falcon, NY).
- Cell strainer 70 μm (BD Falcon, NY).
- Liquid N_2 storage tank.
- Microscope DIAPHOT 200 (Nikon, Japan).
- 1.5-mL microcentrifuge tubes (Eppendorf, Germany).
- Bench-top high-speed centrifuge 5810 R (Eppendorf, Germany).
- Anesthesia machine (Vet Tech Solutions, UK).

- Eppendorf Safe-Lock microcentrifuge 1.5-mL tubes (Eppendorf, Germany).
- Sterile plastic disposable pipettes, 5/10/25-mL (Falcon, NY)

6.2.3 Reagent Preparation

Medium
DMEM supplemented with:

1. 10% FBS
2. 4 mM l-glutamine
3. 100 U/mL penicillin, and 100 μg/mL streptomycin.

Phosphate-Buffered Solution (×10 PBS, pH 7.4)
Dissolve the following in MilliQ water to prepare 1 L:

1. KCl 2.0 g
2. KH 2 PO 4 2.45 g
3. Na 2 HPO 4 14.4 g
4. NaCl 80 g

Just after it is dissolved, adjust the pH to 7.4 and fill up to 1000 mL with MilliQ water. Tenfold dilution of the solution with MilliQ water will provide normal PBS.

Dissociation Buffer
The dissociation solution is made of PBS containing:

1. 0.25% trypsin
2. 0.1% collagenase
3. 20% KSR
4. 1 mM $CaCl_2$

6.3 Methods

6.3.1 Preparation of Primary Cells from a Tumor Tissue

In this section, we tried to describe the preparation of single-cell suspensions from tumor tissues. Several parameters should be optimized prior to using the following protocol such as choice of enzymes and total dissociation time (Fig. 6.1). Perform every step under a sterile condition in a laminar flow hood.

1. Excise the tumor tissue from the host body.
2. Submerge the tissue in PBS in sterile conical tube(s) at room temperature as soon as possible.

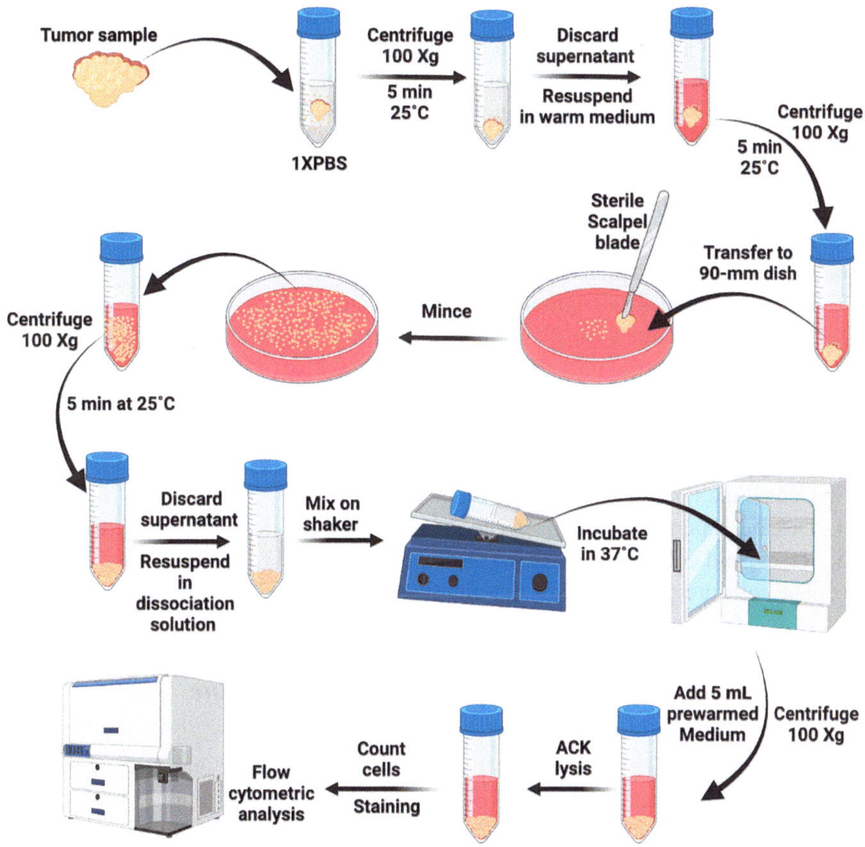

Fig. 6.1 This flow chart illustrates the preparation of cells dissociated from a tumor tissue for flowcytometric analysis. Soon after surgical resection, the tumor is transferred into PBS. Then, the cells in the tissues are enzymatically dissociated for at least 1 h with 0.25% trypsin and 0.1% collagenase. As the final step, the dissociated cells should be treated with ACK lysis buffer to remove red blood cell contamination before proceeding to further experiments

3. Transfer the tissue to a laboratory facility for cell preparation.
4. Centrifuge the tubes with tumor tissue at 100 × g for 5 min at 25 °C.
5. Carefully discard the supernatant using a sterile pipette.
6. Resuspend the tumor tissue in 5 mL of medium warmed at 37 °C.
7. Centrifuge the tubes with tumor tissue at 100 × g for 5 min at 25 °C.
8. Discard the supernatant which contains dead cells carefully using a sterile pipette.
9. Transfer the tumor tissue to a 90-mm tissue culture dish.
10. Mince the tissue into small pieces ranging from 1 to 3 mm^3 with a sterile scalpel in a warm medium.

11. Transfer the minced tumor tissue in pre-warmed medium into 15-mL conical tubes using a micropipette with a sterile 1000-μL tip whose point is cut to get a suitable size of hole for the minced tissue to get inside.
12. Centrifuge the tubes at 100 × g for 5 min at 25 °C.
13. Discard supernatant with a sterile Pasteur pipette.
14. Add ~2 mL of pre-warmed medium.
15. Add 2 mL of the dissociation solution containing 0.25% trypsin, 0.1% collagenase, 20% KSR, and 1 mM $CaCl_2$.
16. Put the tubes on a shaker at 37 °C for at least 1 h.
17. Pick up the tubes and spin down the dissociated cells.
18. Add equal volume (~4 mL) of pre-warmed medium containing 10% FBS.
19. Centrifuge tubes at 100 × g for 5 min at 25 °C.
20. Discard supernatant with a sterile Pasteur pipette.
21. Add 2 mL of ACK lysis buffer and incubate at 25 °C for 1 min.
22. Add 2 mL of pre-warmed DMEM.
23. Centrifuge tubes at 100 × g for 5 min at 25 °C.
24. Discard supernatant with a sterile Pasteur pipette without disturbing the pellet.
25. Resuspend cells in pre-warmed DMEM medium.
26. The cells are ready for flow cytometric analysis.

6.3.2 *Preparation of Cells from a Cancer Cell Line*

Since cryopreservation particularly makes the cells vulnerable and delicate to damage such as by vortex and/or centrifugation at high gravity, tapping with fingers and/or centrifugation at low speed whenever it is necessary is highly recommended for the steps in this section.

1. Pick up the vial of cancer cells from liquid nitrogen storage using metal forceps.
2. Immerse the vial in a 37 °C water bath without submerging the cap.
3. Pick up the vial from the water bath when the liquid half defrosts.
4. Sterile the surface of the vial with 70% ethanol and place it in the hood.
5. Take off the cap and gently transfer the thawed cells into a sterile 15-mL conical tube.
6. Slowly add 1 mL of pre-warmed DMEM containing 10% FBS drop by drop.
7. Add further 3 mL of pre-warmed DMEM containing 10% FBS.
8. Centrifuge the tube at 200 × g for 5 min at 25 °C.
9. Aspirate the supernatant without disturbing the cell pellet.
10. Resuspend the cell pellet in a suitable volume of pre-warmed DMEM containing 10% FBS.
11. Seed the cells in a 60-mm dish.
12. Incubate the dish in an incubator at 37 °C under 5% CO_2.
13. When the cells become 80% confluent, passage the cells until the number increases enough for the experiments or other stocks.

6.3.3 *Passage of Cancer Cells*

In the following steps, we will try to passage the cells to be in stable condition because the cells after reviving from the stock may not exhibit stable growth in the regular condition. In that case, the cells often need at least three passages to get stable growth in regular condition.

1. Detach the cells with 0.05% trypsin by incubating at 37 °C for 5 min.
2. Stop the digestion by adding DMEM containing 10% FBS.
3. Transfer the cells to a new sterile 15-mL conical tube.
4. Centrifuge the tube at 200 × g for 5 min at 25 °C.
5. Aspirate the supernatant without disturbing the cell pellet.
6. Resuspend the cells in 5 mL of fresh medium.
7. Count the cells with a hemocytometer using a small aliquot stained with trypan blue.
8. Seed the cells in dishes as many as you need at 0.3×10^6 cells/60-mm dish.
9. Incubate the dishes at 37 °C with 5% CO_2.
10. After 5 days, the cells will be ready (Fig. 6.2) for isolating CSCs cells using FACS.

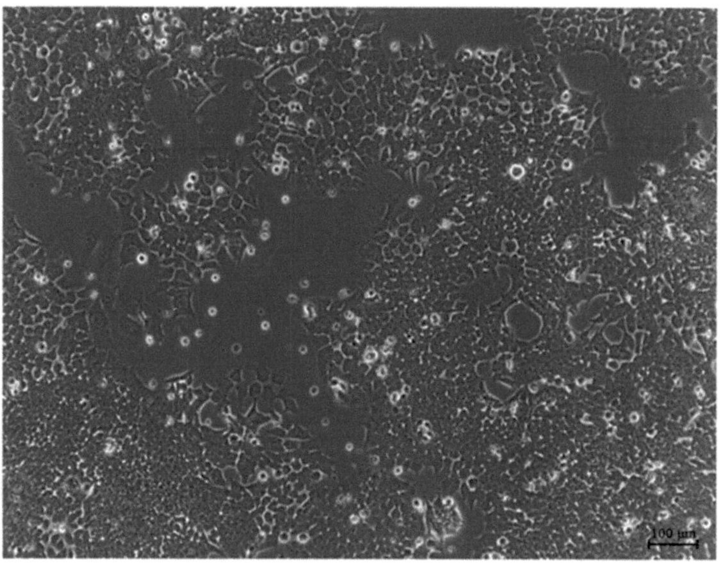

Fig. 6.2 A photograph of HCT116 cells taken after three passages. Scale bar = 100 μm

6.3.4 Identification of CSCs by Cell Surface Markers

In this section, cells are immuno-stained to detect the cells expressing CSC markers such as CD44 and CD133. Using the flow cytometric analysis, researchers can identify the population of CSCs in each tumor.

1. Detach the cells with 0.05% trypsin by incubating at 37 °C for 5 min.
2. Stop the digestion by adding DMEM containing 10% FBS.
3. Transfer the cells to a new sterile 15-mL conical tube.
4. Centrifuge the tube at 200 × g for 5 min at 25 °C.
5. Aspirate the supernatant without disturbing the cell pellet.
6. Resuspend the cells in PBS supplemented with 1% BSA by gently pipetting and spin down the cells by brief centrifugation at 25 °C.
7. Repeat Step 6 three times to wash the cells.
8. Resuspend the cells in PBS supplemented with 1% BSA and filter the suspension through a 75-μm nylon mesh filter to eliminate clumps and debris.
9. Collect the cell suspension in a 15-mL conical tube.
10. Count the cells with a hemocytometer or automated cell counting system using a small aliquot of cells stained with trypan blue.
11. Prepare aliquots of cells as many as required in 1.5-mL tubes containing 1×10^6 cells per tube.
12. Centrifuge tubes at 200 × g for 5 min at 4 °C.
13. Aspirate the supernatant and resuspend the cells in cold PBS containing paraformaldehyde at a final concentration of 4%.
14. Incubate the cells for 20 min at 4 °C to fix the cells.
15. Spin down the cells by brief centrifugation at 4 °C and resuspend the cell pellet with cold PBS.
16. Repeat Step 15 three times.
17. Adjust the cell concentration to 10^7 cells/mL with cold PBS containing 1% BSA and proceed to the next step.

Direct Detection by Antibody Conjugated with Fluorescence

18. Add fluorescence labelled antibodies against the targeted maker (e.g., anti-CD44-PE, anti-CD133-APC, etc.) (Note 1).
19. Incubate the tubes on ice or at 4 °C for 30 min in the dark.
20. Spin down the cells by brief centrifugation at 4 °C and resuspend the cell pellet with cold PBS containing 1% BSA.
21. Repeat Step 20 three times.
22. Adjust the cell concentrations between the tubes (e.g., 1 × 106 cells/100μL) (Note 2).
23. Apply the cells for flow cytometric analysis.

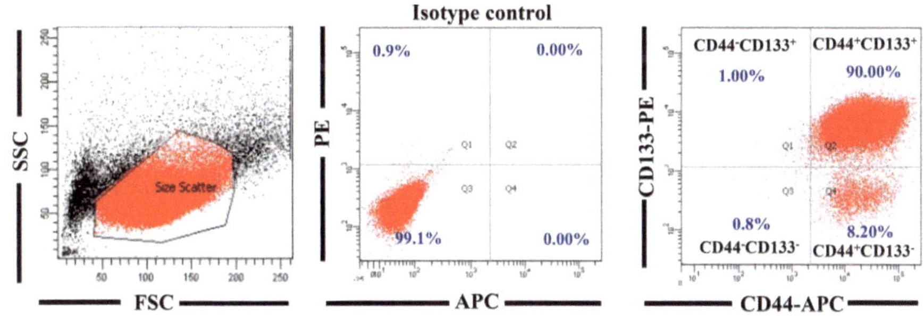

Fig. 6.3 Flow cytometry analysis of the xenografted tumor cells derived from HCT116 cells. Direct staining with anti-human CD44 antibody conjugated with APC and anti-human CD133 antibody conjugated with PE. The CSC subpopulation appears in Q2 dimension as $CD44^{+}/CD133^{+}$

Indirect Detection by the Primary and the Secondary Antibodies

24. Add the primary antibody and incubate for 40 min at 4 °C (Note 3).
25. Centrifuge the tubes at 200 × g for 5 min at 4 °C and resuspend the cells with 100 μL of cold PBS containing 1% BSA.
26. Repeat Step 25 three times.
27. Add the fluorescent-labelled secondary antibody specific to the primary antibody (Note 4).
28. Incubate the tubes on ice in the dark for 30 min.
29. Centrifuge the tubes at 200 × g for 5 min at 4 °C and resuspend the cells with 100 μL of cold PBS containing 1% BSA.
30. Repeat Step 29 three times.
31. Resuspend cell pellet if some different buffer is recommended for the flow cytometer.
32. Apply the cells for flow cytometric analysis (Fig. 6.3).

6.3.5 Separation of CSC Subpopulation

A technique to isolate some cell populations in antigen-specific manner by flow cytometry is named fluorescence-activated cell sorting (FACS). As compared to a non-sorting analysis, sorting allows more precise and specific characterization of cells as a significant component among heterogeneous populations. On the other hand, it is crucial for subsequent culture to keep the viability of the cells without contamination during the procedure. All the following steps should be performed in sterile conditions.

1. Prepare a cell suspension as in the previous Sects. 6.3.1 or 6.3.2.
2. Adjust the cell concentration to 10^7 cells/mL in FACS buffer in three to four 1.5-mL tubes (Note 5).

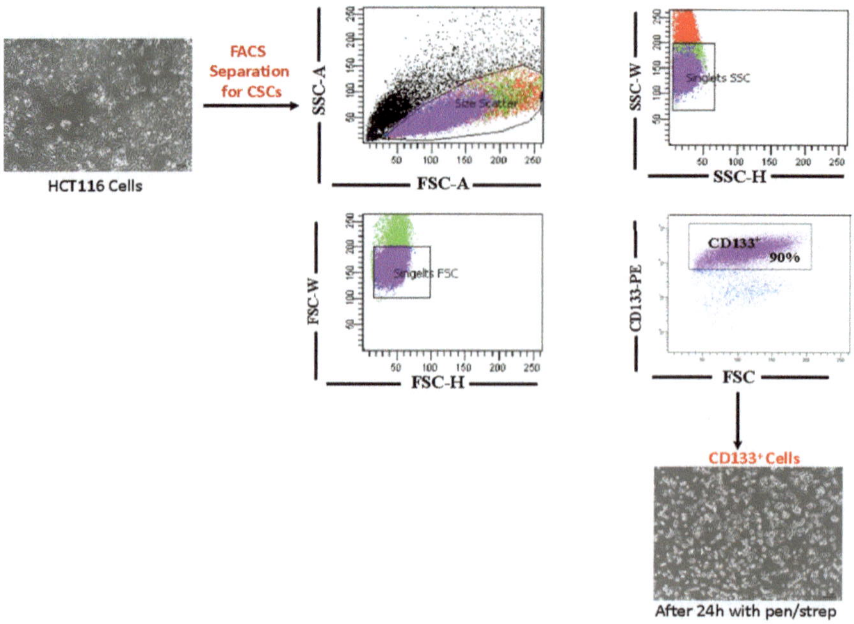

Fig. 6.4 Flow cytometric analysis during FACS. The analysis shows 90% of $CD133^+$ subpopulation is successfully isolated from the bulk of HCT116 cells

3. Add the optimized dilution of fluorophore-conjugated antibodies to the tubes of cell suspension.
4. Incubate on ice for 20 min in the dark. Cf. Avoid longer time incubation.
5. Spin down the cells at 300 × g at 4 °C, aspirate the supernatant leaving 0.1 mL and resuspend the cells with ice-cold FACS buffer.
6. Repeat Step 5 twice.
7. Re-suspend the cells by adding 0.9 mL FACS buffer supplemented with penicillin and streptomycin.
8. Apply the cells to FACS to separate and collect the target cells (Fig. 6.4).

Notes

1. Optimize the dilution following the manufacturer's recommendation or determine in advance. Consider preparing isotype controls and unstained cells to confirm specific staining and autofluorescence.
2. Resuspend cell pellet if some different buffer is recommended for the flow cytometer.
3. Optimal dilution should be followed by the manufacturer's recommendation or determined in advance. Consider preparing isotype controls and cells without antibody treatment.

4. PE-labelled anti-mouse IgG goat antibody, APC-labelled anti-rabbit IgG sheep antibody, etc. Optimal dilution should be followed by the manufacturer's recommendation or determined in advance. Consider preparing the cells without the primary and the secondary antibody treatment for autofluorescence.
5. Consider the cells for isotype controls and without fluorescent dye.

References

Afify SM, Seno M. Conversion of stem cells to cancer stem cells: undercurrent of cancer initiation. Cancers (Basel). 2019;11(3):345.

Afify SM, Hassan G, Seno A, Seno M. Cancer-inducing niche: the force of chronic inflammation. Br J Cancer. 2022;127(2):193–201.

Ayob AZ, Ramasamy TS. Cancer stem cells as key drivers of tumour progression. J Biomed Sci. 2018;25(1):20.

Bonnet D, Dick JE. Human acute myeloid leukemia is organized as a hierarchy that originates from a primitive hematopoietic cell. Nat Med. 1997;3:730–7. https://doi.org/10.1038/nm0797-730.

Goodwin J, Laslett AL, Rugg-Gunn PJ. The application of cell surface markers to demarcate distinct human pluripotent states. Exp Cell Res. 2020;387(1):111749.

Hsu CC, Chiang CW, Cheng HC, Chang WT, Chou CY, Tsai HW, Lee CT, Wu ZH, Lee TY, Chao A, Chow NH, Ho CL. Identifying LRRC16B as an oncofetal gene with transforming enhancing capability using a combined bioinformatics and experimental approach. Oncogene. 2011;30:654–67.

Jaggupilli A, Elkord E. Significance of CD44 and CD24 as cancer stem cell markers: an enduring ambiguity. Clin Dev Immunol. 2012;2012:708036.

Karsten U, Goletz S. What makes cancer stem cell markers different? Springerplus. 2013;2(1):301.

Kim WT, Ryu CJ. Cancer stem cell surface markers on normal stem cells. BMB Rep. 2017;50(6): 285–98.

Walcher L, Kistenmacher AK, Suo H, Kitte R, Dluczek S, Strauß A, Blaudszun AR, Yevsa T, Fricke S, Kossatz-Boehlert U. Cancer stem cells-origins and biomarkers: perspectives for targeted personalized therapies. Front Immunol. 2020;11:1280.

Chapter 7
Enrichment of Cancer Stem Cell from Malignant Tumor

Abstract A better understanding of cancer stem cells may facilitate the prevention and treatment of cancers. The most critical reason for treatment failure is the presence of cancer stem cells, which are generally present as the small population in cancer tissues. Therefore, the enrichment and isolation of these subpopulations will be a great help for the progress of cancer research. Currently many methods for CSC enrichment are proposed, including those based on stem cell surface markers, suspension culture, the mitochondrial membrane potential, cell division, and resistance to cytotoxic compounds or hypoxia. CD44 is a hyaluronic acid receptor present on cell surface as a type I transmembrane glycoprotein. CD44 is well known as a marker for cancer stem cells regulating the microenvironment and metastasis of tumors.

In this chapter we will focus on the procedure to enrich the subpopulation expressing CD44 with hyaluronic acid. Furthermore, we will describe how to assess the effect of hyaluronic acid on the activation of principal pluripotency genes such as OCT3/4, SOX2, KLF4, and Nanog, and how to confirm the enrichment of cancer stem cells comparing them to parent cells.

Keywords Cancer stem cells · CD44 · Tumor microenvironment · CSC enrichment

7.1 Introduction

In cancer tissues, cancer stem cells are generally present in a small number and are in large part responsible for treatment failure (Afify et al. 2022). Being a multifaceted transmembrane glycoprotein, a hyaluronic acid (HA) receptor CD44 has been recognized as a marker for CSCs. Several studies have suggested that CD44 was involved in tumor invasion and epithelial-mesenchymal transition (EMT) (Toole 2010).

Biosynthesis of hyaluronic acid (HA) in tumor microenvironments has been shown to promote cancer aggressiveness and poor clinical outcomes that adversely affect overall survival rates (Tammi et al. 2008; Auvinen et al. 2013; Chen et al. 2018). In the microenvironment of the brain tumor and metastasis, HA and CD44 interact cooperatively (Pibuel et al. 2021). Meanwhile, we found CD44 was

S. M. Afify, M. Seno, *Methods in Cancer Stem Cell Biology*,
https://doi.org/10.1007/978-981-99-1331-2_7

overexpressed in nine cell lines derived from human glioma (Sugii et al. 2016). In glioma patients, CD44 may be crucial to prognosis. This result suggested that these cell lines should contain the subpopulation of CSCs expressing CD44.

Enrichment of the CSCs would allow more efficient drug screening targeting CSCs themselves. Since CD44 is a receptor for HA, HA could be exploited to enrich the CD44 expressing CSCs in spheroid culture in an inexpensive manner.

Brain tumors have high mortality and morbidity rates due to their localization and invasive nature (Gavrilovic and Posner 2005). Most brain tumor deaths are caused by glioblastoma (GBM), which account for almost 30% of all primary brain tumors of which 80% are malignant. Invading into the surrounding brain tissue rather than metastasizing, GBM is defined as a diffuse glioma, which is characterized by a high aptitude to infiltrate into surrounding brain tissue (Ceccarelli et al. 2016).

Typically, CD44, SOX2, OCT4, NANOG, CD133, and ABCG2 are well known as the markers of GBM CSC (Islam et al. 2015; Lathia et al. 2015; Bradshaw et al. 2016). CD44 inhibitions have been suggested as therapeutic agents against GBM CSC although it is still controversial (Breyer et al. 2000). Therefore, conventionally simple isolation methods of CSCs are currently anticipated. We show in this chapter how to enrich CD44-expressing cells following our previous publication (Vaidyanath et al. 2017).

7.2 Materials

7.2.1 Reagents

- Human brain glioblastoma cell line U-251 MG cell (ECACC, UK).
- Dulbecco's Modified Eagle Medium (DMEM) (Thermo Fisher Scientific, MA).
- Minimum essential media (MEM) non-essential amino acid solution 100× (Wako-Fujifilm, Japan).
- Fetal bovine serum (FBS) (Gibco Life Technologies, MA).
- Penicillin/streptomycin mixed solution (Nacalai Tesque, Japan).
- L-glutamine (Nacalai Tesque, Japan).
- 2-Mercaptoethanol (Sigma-Aldrich, MO).
- Trypsin-EDTA solution. 0.25% trypsin (Sigma-Aldrich, MO).
- Phosphate-buffered saline (PBS) (Genesee Scientific, CA).
- Hank's balanced salt solution (HBSS) (Genesee Scientific, CA).
- Ethanol (Sigma-Aldrich, MO).
- Blocking buffer: HBSS containing 1% BSA.
- FACS buffer: in HBSS containing 0.2% BSA.
- $CaCl_2$ (Sigma-Aldrich, MO).
- Dimethyl sulfoxide (DMSO) (Sigma-Aldrich, MO).
- Appropriate tissue culture plates and supplies.
- Povidone iodine scrub solution 7.5% iodine (Dynarex, NY).
- Isoflurane (WAKO-Fujifilm, Japan).

- BALB/c-nu/nu immunodeficient mice, 4-week-old female (Charles River laboratories, Japan).

7.2.2 Equipment

- Syringe needles, 24G.
- 37 °C water bath.
- Cell counting instrument (e.g., Vi-Cell™ or hemocytometer).
- Sterile cotton buds.
- Iris/eye scissors, straight.
- Operating scissors.
- Standard forceps, 12–13 cm, straight, pointed.
- CO_2 incubator MCO-19AIC(UV) (Panasonic, Japan).
- Bio-Clean Bench CCV-1307E (Hitachi, Japan).
- Olympus IX81 inverted microscope (Olympus, Japan).
- Laser scanning confocal microscope FV-1000 (Olympus, Japan).
- Tissue culture-treated plate, 60-mm dish (TPP Techno Plastic Products AG, Switzerland).
- Tissue culture-treated plate, 100-mm dish (TPP Techno Plastic Products AG, Switzerland).
- Falcon® Conical 15-mL centrifuge tubes (BD Falcon, NY).
- Falcon® Conical 50-mL centrifuge tubes (BD Falcon, NY).
- Cell Strainer 70 μm (BD Falcon, NY)
- Liquid N_2 storage tank.
- Microscope DIAPHOT 200 (Nikon, Japan)
- 1.5-mL microcentrifuge tubes (Eppendorf, Germany).
- Bench-top high-speed centrifuge 5810 R (Eppendorf, Germany).
- Anesthesia machine (Vet Tech Solutions, UK).
- Eppendorf Safe-Lock microcentrifuge 1.5-mL tubes (Eppendorf, Germany).
- Sterile plastic disposable pipettes, 5/10/25-mL (Falcon, NY)

7.2.3 Reagent Preparation

Normal Medium
DMEM supplemented with:

1. 10% FBS
2. 4 mM l-glutamine
3. 100 U/mL penicillin and 100 μg/mL streptomycin

Enrichment Medium
DMEM supplemented with:

1. 100 μg/mL of HA
2. 10% FBS
3. 4 mM l-glutamine
4. 100 U/mL penicillin, and 100 μg/mL streptomycin

Sphere Medium
DMEM supplemented with:

1. 100 μg/mL HA
2. Insulin/Transferrin/Selenium-X (ITS-X)
3. 4 mM l-glutamine
4. 100 U/mL penicillin, and 100 μg/mL streptomycin

Phosphate-Buffered Solution (×10 PBS, pH 7.4)
Dissolve the following in MilliQ water to prepare 1 L.

1. KCl 2.0 g
2. KH 2 PO 4 2.45 g
3. Na 2 HPO 4 14.4 g
4. NaCl 80 g

Just after it is dissolved, adjust the pH to 7.4 and fill up to 1000 mL with MilliQ water. Tenfold dilution of the solution with MilliQ water will provide normal PBS.

Dissociation Buffer
The dissociation solution is made of PBS containing:

1. 0.25% trypsin
2. 0.1% collagenase
3. 20% KSR
4. 1 mM $CaCl_2$

7.3 Methods

7.3.1 Enrichment CD44 Expressing Subpopulation

We are trying to concentrate the cell population expressing CD44 in the presence of HA. In this process HA will link the CD44 positive cells one another to form colonies or spheroids. As a result, CD44 positive CSC subpopulation could be enriched.

7.3.1.1 Culture-Adherent Culture in Presence of Hyaluronic Acid (HA)

1. Prepare the cells in a 60-mm dish under adherent condition without HA at a density no more than 80%.
2. Warm 0.05% trypsin-EDTA and other media up to 37 °C in a water bath.
3. Aspirate the culture media from the 80%-confluent dishes.
4. Rinse the dish with trypsin-EDTA twice.
5. Add 2 mL of trypsin-EDTA per dish and incubate the dishes for 3–5 min at 37 ° C in a CO_2 incubator.
6. Confirm the cell to get detached from the bottom.
7. Add equal volume of serum-containing normal medium.
8. Mix well by pipetting.
9. Transfer the suspension to a sterile 15-mL conical centrifuge tube.
10. Centrifuge the cell suspension at 120 × g for 5 min at 25 °C.
11. Aspirate the medium without disturbing the pellet.
12. Resuspend the cells in an enrichment medium and count cell number.
13. Adjust the concentration of the cell at 5 cells/mL with enrichment media.
14. Seed the cells 100 μL/well of a 96-well plate. Cf., 0.5 cell/well. Prepare 10–20 96-well plates for adherent culture.
15. Keep the culture until the colony becomes visible (Fig. 7.1) (Note 1).
16. Transfer the cells to larger sizes of wells when the colonies become visible (Note 2).
17. Proceed to spheroid culture when the cell number becomes enough (Note 3).

7.3.1.2 Suspension Culture

1. Aspirate culture medium from the colony-forming cells in a dish at 80% confluence.
2. Rinse the dish twice with PBS to remove dead cells.
3. Trypsinize the cells with trypsin-EDTA by incubating the dishes at 37 °C in a CO_2 incubator (Note 4).
4. Collect the colony-forming cells with a sterile Pasteur pipette from the dish and transfer into a sterile 15-mL conical tube.
5. Add sufficient trypsin-EDTA to the tube up to approximately 2 mL if the amount is too small.
6. Incubate the tube at 37 °C until the cells get apart from the colonies (Note 5).
7. Examine under a microscope to determine if the colonies are digested and singular cells are visible.
8. Add 5 mL pre-warmed enrichment medium which contains 10% FBS.
9. Centrifuge the tube for 5 min at 120 × *g* at 25 °C.
10. Aspirate the supernatant without disturbing the pellet.
11. Resuspend the cells with pre-warmed sphere medium and centrifuge for 5 min at 120 × *g* at 25 °C.

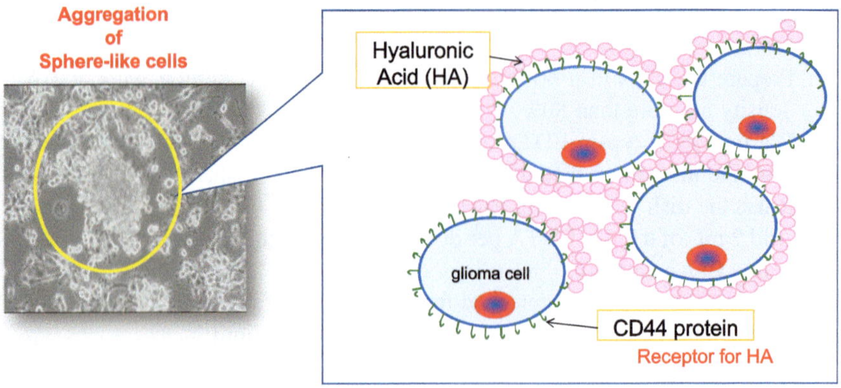

Fig. 7.1 Representative images for enrichment of cells expressing CD44

12. Repeat Step 11.
13. Aspirate supernatant without disturbing cell pellet.
14. Gently resuspend cells in a fresh sphere medium.
15. Count the cell number and adjust the cell concentration 500 cells/mL with sphere medium.
16. Seed 1000 single dissociated cells per well of a 6-well ultralow attachment plate in 2 mL of sphere medium.
17. Observe cell cultures daily under a microscope.
18. Feed the sphere medium every 3–4 days after the first week (Note 6).

7.3.1.3 Passaging Spheres

1. After 2 weeks, collect the spheres by centrifugation at 120 × g for 5 min at 25 °C.
2. Carefully discard the supernatant without disturbing the pellet (Note 7).
3. Trypsinize spheres for 5 min at 37 °C using 0.05 mg/mL trypsin-EDTA solution.
4. Stop trypsinization with serum containing culture media.
5. Transfer the cell suspension to a sterile 15-mL conical tube.
6. Spin down the cells at 120 × g for 5 min at 25 °C.
7. Resuspend the cells in sphere medium.
8. Slowly pipette the cells to prepare to dissociate into single cells.
9. Filter through a 75-mm mesh (Note 8).
10. Plate cells slightly less than those for the primary sphere formation culture (Fig. 7.2).

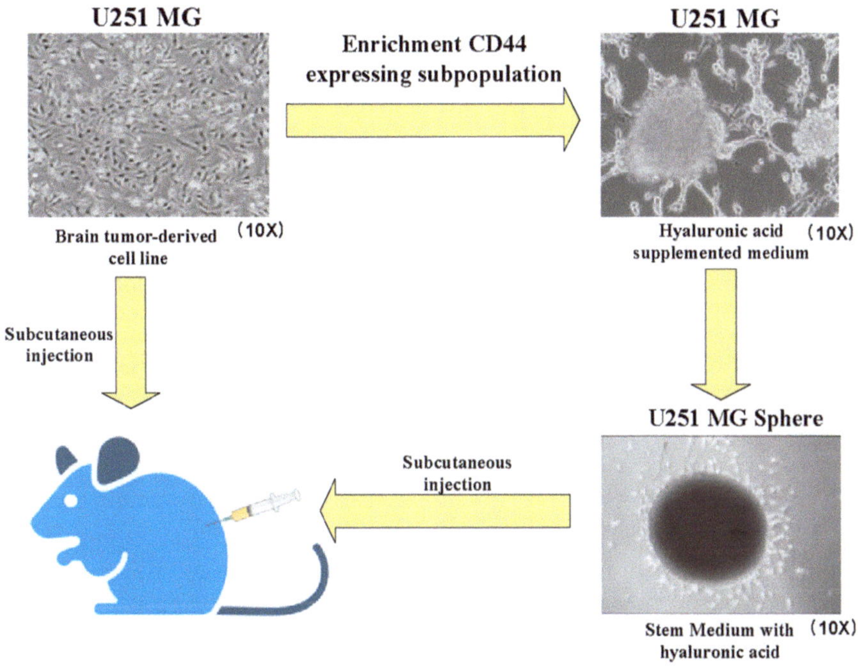

Fig. 7.2 Scheme of CSC establishment with HA. Cancer cells were cultured in an adherent plate until 80% confluent. Cells were then transferred into culture condition supplemented with HA. 1×10^4 cells/mL were transferred into low attachment dishes for sphere culture in the presence of HA. The formed spheroids were injected into a nude mouse, where they generated a malignant tumor. The primary cells from the excised tumor were cultured in an adherent dish until they became confluent

7.3.2 *Cell Preparation for Injection*

1. Collect spheres from culture dishes and transfer them to a sterile 15-mL conical tube.
2. Spin down the spheres at 120 × g for 5 min at 25 °C.
3. Aspirate supernatant.
4. Wash spheres with sterile PBS.
5. Spin down the spheres at 120 × g for 5 min at 25 °C.
6. Trypsinize spheres for 5 min at 37 °C using 0.05% trypsin-EDTA solution (Note 9).
7. Stop trypsinization by adding a pre-warmed medium containing 10% FBS.
8. Spin down the cells at 120 × g for 5 min at 25 °C.
9. Discard supernatant without disturbing the pellet.
10. Resuspend pellet of single cells in ice-cold HBSS.
11. Count the cells.
12. Make 1×10^6 cells/0.1mL in 1.5 mL Eppendorf.

13. Keep the tube on ice (Note: Untreated cells should be used as the control).
14. Prepare nude mice for subcutaneous injection.
15. Before injection, clean mice skin with 75% ethanol (Note 10).
16. Invert the syringe to ensure the cells are in suspension.
17. Insert the needle through the base of the tented skin with your dominant hand.
18. Inject the cells subcutaneously.
19. Observe whether any complications occur after returning the animal to its cage.
20. Measure the size and volume of tumors (Note 11).
21. Four weeks later, a malignant tumor should be observed (Fig. 7.3) (Note 12).

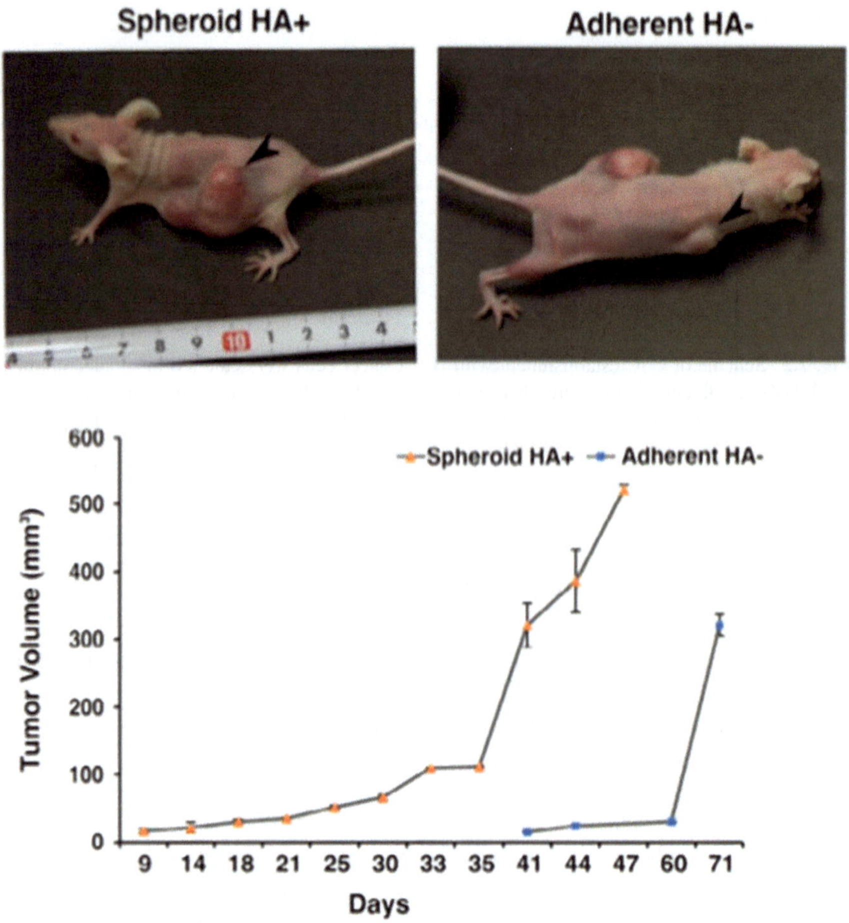

Fig. 7.3 Tumor developed from spheroids of U251MG cells treated with HA and adherent culture of U251MG cells without the HA treatment. The figure copied from our article Vaidyanath et al. (2017)

Notes

1. Generally it takes 1 month or more.
2. A 24-well and a 12-well plate followed by a 35-mm and a 60-mm dish will be necessary to keep culturing to obtain enough cells. Do not skip from a 96-well plate to a 60-mm dish at one time so that the cell growth does not get too slow.
3. It is recommended to make stocks of the cells in LN_2.
4. Confirm the colonies detaching from the dishes.
5. Long incubation in the presence of trypsin affects the cell's viability.
6. Medium exchange should carefully be made so that the spheres may not be lost. Leave approximately 400 μL of medium and add 1.6 mL of fresh medium, without disturbing or disrupting the developing spheres.
7. Note: spheres could be socked in LN2 as the sphere forming cells in HA medium.
8. This step eliminates aggregated cells.
9. Keep monitoring the sphere during trypsinization and do not incubate for a long time with trypsin because long incubation affects cell viability.
10. Anesthesia is not necessary but makes handling easier.
11. Measurements were made every 3–4 days. Follow the formula (0.5 × width) 2 × length, whereas "width" is the smallest diameter and "length" is the longest diameter, to calculate volume.
12. After 4 weeks a tumor derived from a spheroid treated with HA significantly grew faster than that derived from untreated cells.

Acknowledgment We would like to thank all our lab members who wrote the main paper in the journal of stem cell research and therapy, Vaidyanath A., Mahmud H., Apriliana C.K., Oo A.K.K., Seno A., Asakura M., Kasai T., Seno M.

References

Afify SM, Hassan G, Seno A, Seno M. Cancer-inducing niche: the force of chronic inflammation. Br J Cancer. 2022;127(2):193–201.

Auvinen P, Rilla K, Tumelius R, Tammi M, Sironen R, et al. Hyaluronan synthases (HAS1–3) in stromal and malignant cells correlate with breast cancer grade and predict patient survival. Breast Cancer Res Treat. 2013;2:277–86.

Bradshaw A, Wickremsekera A, Tan ST, Peng L, Davis PF, Itinteang T. Cancer stem cell hierarchy in glioblastoma multiforme. Front Surg. 2016;3:21.

Breyer R, Hussein S, Radu DL, et al. Disruption of intracerebral progression of C6 rat glioblastoma by in vivo treatment with anti-CD44 monoclonal antibody. J Neurosurg. 2000;92(1):140–9.

Ceccarelli M, Barthel FP, Malta TM, Sabedot TS, Salama SR, Murray BA, Morozova O, Newton Y, Radenbaugh A, Pagnotta SM, et al. Molecular profiling reveals biologically discrete subsets and pathways of progression in diffuse glioma. Cell. 2016;164:550–63.

Chen JE, Pedron S, Shyu P, Hu Y, Sarkaria JN, Harley BAC. Influence of hyaluronic acid transitions in tumor microenvironment on glioblastoma malignancy and invasive behavior. Front Mater. 2018;5:39.

Gavrilovic IT, Posner JB. Brain metastases: epidemiology and pathophysiology. J Neuro-Oncol. 2005;75:5–14.

Islam F, Gopalan V, Smith RA, Lam AKY. Translational potential of cancer stem cells: A review of the detection of cancer stem cells and their roles in cancer recurrence and cancer treatment. Exp Cell Res. 2015;335(1):135–47.

Lathia JD, Mack SC, Mulkearns-Hubert EE, Valentim CLL, Rich JN. Cancer stem cells in glioblastoma. Genes Dev. 2015;29(12):1203–17.

Pibuel MA, Poodts D, Díaz M, Hajos SE, Lompardía SL. The scrambled story between hyaluronan and glioblastoma. J Biol Chem. 2021;296:100549.

Sugii Y, Kasai T, Ikeda M, Vaidyanath A, Kumon K, Mizutani A, Seno A, Tokutaka H, Kudoh T, Seno M. A unique procedure to identify cell surface markers through a spherical self-organizing map applied to DNA microarray analysis. Biomark Cancer. 2016;8:17–23.

Tammi RH, Kultti A, Kosma VM, Pirinen R, Auvinen P, et al. Hyaluronan in human tumors: pathobiological and prognostic messages from cell-associated and stromal hyaluronan. Semin Cancer Biol. 2008;4:288–95.

Toole BP. Hyaluronan-CD44 interactions in cancer: paradoxes and possibilities. Clin Cancer Res. 2010;15(24):7462–8.

Vaidyanath A, Mahmud H, Apriliana CK, Oo AKK, Seno A, Asakura M, Kasai T, Seno M. Hyaluronic acid mediated enrichment of cd44 expressing glioblastoma stem cells in u251mg xenograft mouse model. J Stem Cell Res Ther. 2017;7:384. https://doi.org/10.4172/2157-7633.1000384.

Chapter 8
Isolation of Single Clonal Cell from Primary Cultured Cells and Establishment of a Cancer Stem Cell Line

Abstract In the last 25 years, cancer research has shed light on cancer stem cells since the leukemic cancer stem cells were first identified. Till now, the characterizations of cancer stem cells are still not enough and need more investigation. However, isolation of cancer stem cells is very hard because the population of cancer stem cells in cancer tissues are very small. Even once isolated, the establishment of cancer stem cell lines is still tough due to the difficulty of keeping phenotypes in vitro. Therefore, many trials have been challenged to isolate and maintain the subpopulation of cancer stem cells.

The protocol outlined in this book chapter describes the method to isolate and characterize the stem-like subpopulation, which is condensed in the presence of hyaluronic acid as $CD44^+$ population, by single cell isolation procedure from human cancer cell lines. In addition, we refer to the propagation method of this subpopulation of cells by successive rounds of sphere formation. These approaches will help scientists establish novel in vitro cancer stem cells, which may play a key role during carcinogenesis.

Keywords Cancer stem cell · $CD44^+$ · sphere formation

8.1 Introduction

Globally, cancer is the leading cause of death (Sung et al. 2021). Many new therapeutic agents and treatments are being developed worldwide. And the competition among pharmaceutical companies is getting fiercer than ever. Surgery, chemotherapy, and radiotherapy are not always possible to completely remove tumor cells. The residual cells will occasionally lead to poor prognosis causing recurrences. The responsible cells are generally attributed to CSCs due to their resistance to chemotherapy and radiation allowing recurrence and metastasis (Phi et al. 2018; Afify and Seno 2019; Afify et al. 2019, 2022; Arnold et al. 2020; Nawara et al. 2020, 2021; Zahra et al. 2022).

Therefore, CSC lines separated from heterogeneous tumor masses would become a fundamental source in cancer research. Various trials have been conducted to establish CSC lines to be applied for the development of advanced cancer treatment.

S. M. Afify, M. Seno, *Methods in Cancer Stem Cell Biology*,
https://doi.org/10.1007/978-981-99-1331-2_8

Human CSCs or cancer initiating cells (CICs) were first isolated from acute myeloid leukemia (AML) as $CD34^{+}/CD38^{-}$ cells (Lapidot et al. 1994). Not only from leukemia but from solid malignant tumors, the isolation of CSCs has been described depending on the cell surface markers such as CD44 and CD133, and on the capability of sphere formation. Multiple studies have consistently shown that CSCs can be isolated from various types of cancer, including gastrointestinal cancers (Haraguchi et al. 2006), glioma (Kondo et al. 2004), lung cancer (Ho et al. 2007; Nakatsugawa et al. 2011), ovar ian cancer (Szotek et al. 2006; Yasuda et al. 2013), thyroid cancer (Mitsutake et al. 2007), renal cell carcinoma (Nishizawa et al. 2012), and malignant lymphoma (Moti et al. 2014) and hepatocellular carcinoma (Chiba et al. 2006).

Although these ways of isolation were effective in preparing some CSC lines, models of CSC from different types of malignant tumors are required to fulfil the development of all the treatments against cancer. The common problem among all of the trials is that the CSC subpopulation in patient-derived samples is too small for the purpose of analyses. The concept of CSC has therefore been uncommon for a long period after the discovery, so that current therapeutic strategies are not so much different from the traditional and conventional approaches.

In this context, enrichment of CSC subpopulation in advance should be critical to isolate and establish CSC lines from tumors. And the enrichment could depend on the cell surface markers such as CD44 exploiting the antibodies or specific ligands to the markers.

In this chapter, we show the process of isolating specific subpopulations of CSC from the primary culture using single colony isolation or FACS.

8.2 Materials

8.2.1 Reagents

- Human brain glioblastoma cell line U-251 MG cell (ECACC, UK).
- Dulbecco's Modified Eagle Medium (DMEM) (Thermo Fisher Scientific, MA).
- Minimum essential media (MEM) non-essential amino acid solution 100× (Wako-Fujifilm, Japan).
- Fetal bovine serum (FBS) (Gibco Life Technologies, MA).
- Penicillin/streptomycin mixed solution (Nacalai Tesque, Japan).
- L-glutamine (Nacalai Tesque, Japan).
- 2-Mercaptoethanol (Sigma-Aldrich, MO).
- Trypsin-EDTA solution. 0.25% trypsin (Sigma-Aldrich, MO).
- Phosphate buffered saline (PBS) (Genesee Scientific, CA).
- Hank's balanced salt solution (HBSS) (Genesee Scientific, CA).
- Ethanol (Sigma-Aldrich, MO).
- Blocking buffer: HBSS containing 1% BSA.
- FACS buffer: in HBSS containing 0.2% BSA.

- $CaCl_2$ (Sigma-Aldrich, MO).
- Dimethyl sulfoxide (DMSO) (Sigma-Aldrich, MO).
- Appropriate tissue culture plates and supplies.
- Povidone Iodine Scrub Solution 7.5% Iodine (Dynarex, NY).
- Isoflurane (WAKO-Fujifilm, Japan).
- BALB/c-nu/nu immunodeficient mice, 4-week-old female (Charles River laboratories, Japan).

8.2.2 *Equipment*

- Syringe needles, 24G.
- 37 °C water bath.
- Cell counting instrument (e.g., Vi-Cell™ or hemocytometer).
- Sterile cotton buds.
- Iris/eye scissors, straight.
- Operating scissors.
- Standard forceps, 12–13 cm, straight, pointed.
- CO_2 incubator MCO-19AIC (UV) (Panasonic, Japan).
- Bio-Clean Bench CCV-1307E (Hitachi, Japan).
- Olympus IX81 inverted microscope (Olympus, Japan).
- Laser scanning confocal microscope FV-1000 (Olympus, Japan).
- Tissue culture-treated plate, 60-mm dish (TPP Techno Plastic Products AG, Switzerland).
- Tissue culture-treated plate, 100-mm dish (TPP Techno Plastic Products AG, Switzerland).
- Falcon® Conical 15-mL centrifuge tubes (BD Falcon, NY).
- Falcon® Conical 50-mL centrifuge tubes (BD Falcon, NY).
- Cell Strainer 70 μm (BD Falcon, NY).
- Liquid N_2 storage tank.
- Microscope DIAPHOT 200 (Nikon, Japan).
- 1.5-mL microcentrifuge tubes (Eppendorf, Germany).
- Bench-top high-speed centrifuge 5810 R (Eppendorf, Germany).
- Anesthesia machine (Vet Tech Solutions, UK).
- Eppendorf Safe-Lock microcentrifuge 1.5-mL tubes (Eppendorf, Germany).
- Sterile plastic disposable pipettes, 5/10/25-mL (Falcon, NY).

8.2.3 *Reagent Preparation*

Normal Medium
DMEM supplemented with:

1. 10% FBS.
2. 4 mM l-glutamine.
3. 100 U/mL penicillin and 100 μg/mL streptomycin.

Enrichment Medium
DMEM supplemented with:

1. 100 μg/ml of HA.
2. 10% FBS.
3. 4 mM l-glutamine.
4. 100 U/mL penicillin, and 100 μg/mL streptomycin.

Sphere Medium
DMEM supplemented with:

1. 100 μg/ml HA.
2. Insulin/transferrin/selenium-X (ITS-X).
3. 4 mM l-glutamine.
4. 100 U/mL penicillin, and 100 μg/mL streptomycin.

Phosphate-Buffered Solution (×10 PBS, pH 7.4)
Dissolve the following in MilliQ water to prepare 1 l:

1. KCl 2.0 g.
2. KH 2 PO 4 2.45 g.
3. Na 2 HPO 4 14.4 g.
4. NaCl 80 g.

Just after it is dissolved, adjust the pH to 7.4 and fill up to 1000 mL with MilliQ water. Tenfold dilution of the solution with MilliQ water will provide normal PBS.

Dissociation Buffer
The dissociation solution is made of PBS containing:

1. 0.25% trypsin
2. 0.1% collagenase
3. 20% KSR
4. 1 mM $CaCl_2$.

8.3 Methods

8.3.1 Primary Culture of the Malignant Tumor Derived from U-251MG Spheroids (Continued from Chap. 7)

1. Place the tissue excised from mouse xenograft in a sterile 60-mm culture dish.
2. Mince the tumor tissue into small pieces with sterile forceps and scissors.

3. Transfer the tissue pieces in a sterile 50-mL conical centrifuge tube.
4. Add ice-cold HBSS and rinse the pieces vigorously.
5. Centrifuge the suspension at 120 ×g for 5 min at 4 °C.
6. Aspirate the supernatant without disturbing the pellet.
7. Suspend the pellet in 2 mL of ice-cold HBSS.
8. Add 2 mL dissociation buffer.
9. Incubate the tube at 37 °C for at least 1 h with gentle shaking.
10. Add the equal volume of normal DMEM to stop the reaction.
11. Pass the digested tissue through a sterile 70-μm cell strainer (or sterile 75-μm nylon mesh) into a new sterile 50-mL centrifuge tube.
12. Centrifuge the sieved cells at 120 ×g for 5 min at 25 °C.
13. Resuspend the pellet in normal DMEM.
14. Count the cell number.
15. Seed the cells at 0.3×10^6 cells/60-mm dish.
16. Change the medium to fresh normal DMEM after 24 h.
17. Keep the cells growing until 80% confluence before passage (Note 1).

8.3.2 CSC Isolation

8.3.2.1 Fluorescence-Activated Cell Sorting (FACS)

As discussed in Chap. 5 that subpopulation of cells could be sorted alive based on fluorescent labelling. The cells expressing CD44 could be sorted as follows.

1. Start with U-251 MG P1 cells in a 60-mm dish at 80% confluence.
2. Detach cells with 0.05% trypsin-EDTA at 37 °C for 5 min.
3. Stop trypsinization by adding normal medium.
4. Transfer the cells to a new sterile 15-mL conical tube.
5. Centrifuge the cells at 120 ×g for 5 min at 25 °C.
6. Aspirate the supernatant without disturbing the pellet.
7. Resuspend the cells in 5 ml of fresh normal medium.
8. Repeat Steps 5–7 three times.
9. Suspend the cells in an enrichment medium and count the cell number.
10. Seed 1.0×10^4 cells in a 60-mm ultra-low attachment dish (Note 2).
11. Incubate the dishes at 37 °C with 5% CO_2 for 10 days to form spheres.
12. Collect spheres in a new sterile 15-mL conical tube.
13. Centrifuge the cells at 120 ×g for 5 min at 25 °C.
14. Rinse the cells with sterile PBS and spin down at 120 ×g for 5 min at 25 °C.
15. Repeat Step 14 twice.
16. Dissociate the spheres with 0.025% trypsin at 37 °C for 5 min (Note 3).
17. Stop the reaction by adding PBS containing 10% FBS.
18. Count the cell number.
19. Adjust the cell concentration to 10^7 cells/mL in PBS containing 1% BSA.

20. Add the optimized dilution of fluorescent conjugated antibody against CD44 to the suspension of dissociated cells.
21. Incubate on ice for 20 min in the dark.
22. Avoid long incubation to maintain the cell viability.
23. Wash the cells twice with ice-cold PBS.
24. Centrifuge the cells at 120×g at 25 °C.
25. Re-suspend cells in 1 mL of FACS buffer supplemented with penicillin/ streptomycin.
26. Prepare two sterile tubes containing medium supplemented with penicillin/ streptomycin (Note 4).
27. Apply the cells to the FACS apparatus and collect the separated cells.

8.3.2.2 Single Cell Cloning

FACS apparatus is not always available. Single cell cloning is a simple and conventional method to separate CSC subpopulation as follows:

1. Start with U-251 MG P1 cells in a 60-mm dish at 80% confluence.
2. Detach cells with 0.05% trypsin-EDTA at 37 °C for 5 min.
3. Stop trypsinization by adding normal medium.
4. Transfer the cells to a new sterile 15-mL conical tube.
5. Centrifuge the cells at 120 ×g for 5 min at 25 °C.
6. Aspirate the supernatant without disturbing the pellet.
7. Resuspend the cells in 5 ml of fresh normal medium.
8. Repeat Steps 5–7 three times.
9. Suspend the cells in an enrichment medium and count the cell number.
10. Seed 1.0×10^4 cells in a 60-mm ultra-low attachment dish.
11. Incubate the dishes at 37 °C with 5% CO_2 for 10 days to form spheres.
12. Collect spheres in a new sterile 15-mL conical tube.
13. Centrifuge the cells at 120 ×g for 5 min at 25 °C.
14. Rinse the cells with sterile PBS and spin down at 120 ×g for 5 min at 25 °C.
15. Repeat Step 14 twice.
16. Dissociate the spheres with 0.025% trypsin at 37 °C for 5 min.
17. Stop the reaction by adding an enrichment medium.
18. Count the cell number.
19. Adjust the cell concentration to 5 cells/mL with sphere medium.
20. Seed the cells derived from spheres into ten 96-well adhesive plates at 0.5 cell/ well.
21. Allow cells to grow as a colony in each well at 37 °C with 5% CO_2 for approximately 1 month or more until the colony becomes visible in several wells.
22. Detach the cells with 50 μL of 0.025% trypsin-EDTA when the cells cover 80% of the bottom of a well.
23. Stop trypsinization by adding 150 μL of pre-warmed enrichment medium (Note 5).

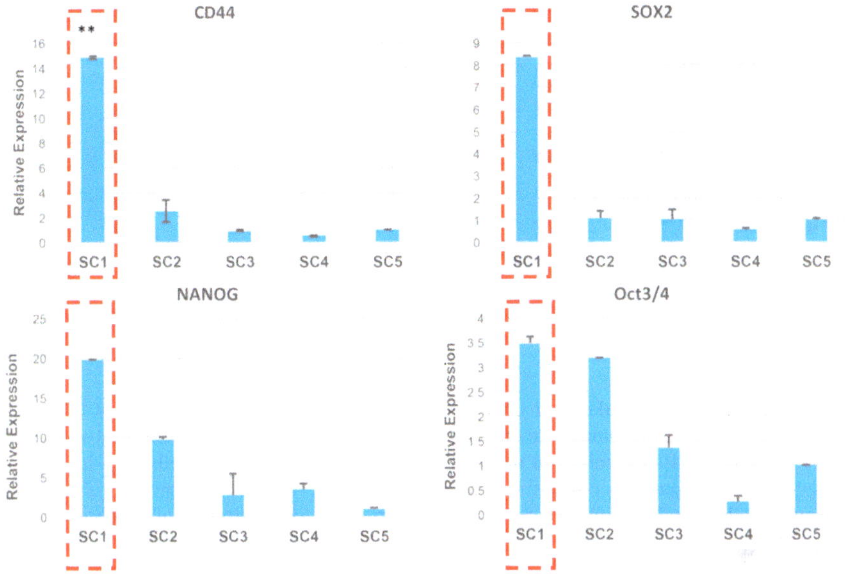

Fig. 8.1 Comparison of stem cell marker expression in five clones obtained by single colony isolation. The results of rt-qPCR were assessed for the expression of CD44, SOX2, NANOG, and OCT3/4 genes. SC1, SC2, SC3, SC4, SC5 indicate the five isolated clones, U251-MGSC1, U-251 MG SC2, U-251-MG-SC3, U-251 MGSC4, and U-251 MG CS5, respectively. The figure is copied from our original paper: Ishii H, Mimura Y, Zahra MH, Katayama S, Hassan G, Afify SM, Seno M. Isolation, and characterization of cancer stem cells derived from human glioblastoma. Am J Cancer Res. 2021 Feb 1;11(2):441–457

24. Transfer the cells into a well of 24-well plate and incubate at 37 °C with 5% CO_2 until the cells cover 80% of the bottom of a well.
25. Scale up the sizes of the plate/dish according to the growth of cells (Note 6).
26. Perform qPCR for CD44 gene expression as well as stemness marker genes. Note: Evaluate the cells from several different wells. The colony enriched cells could show CSC characters (Fig. 8.1).
27. Confirm the expression of CD44 and CD133 by flowcytometry in the isolated one (Fig. 8.2).

8.3.3 *Evaluation of Self-Renewal Potential of the Isolated Cells*

The in vitro limiting dilution assay is performed by counting the frequency of spheres of CSC in suspension culture.

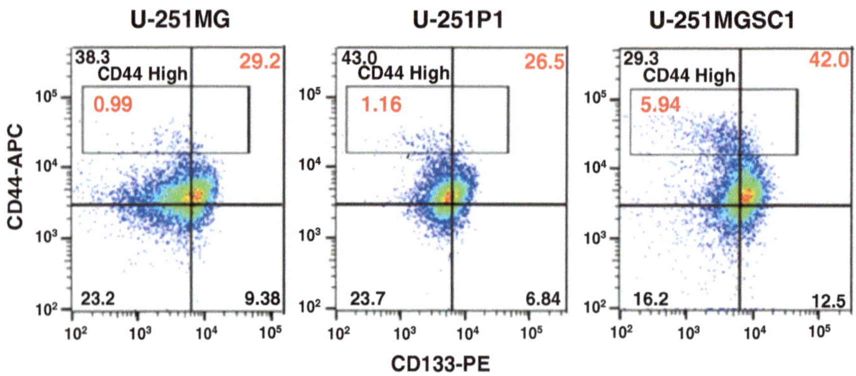

Fig. 8.2 Expression of markers for CSC. The vertical and horizontal axes represent CD44 and CD133, respectively. Red numbers in the top right indicate the percentage of cells positive for both CD44 and CD133. Red numbers in the boxes indicate the percentage of cells with high CD44 expression. The figure is copied from our original paper: Ishii H, Mimura Y, Zahra MH, Katayama S, Hassan G, Afify SM, Seno M. Isolation, and characterization of cancer stem cells derived from human glioblastoma. Am J Cancer Res. 2021 Feb 1;11(2):441–457

1. Grow cells to confluence in a 60-mm dish at 37 °C in an incubator with 5% CO_2.
2. Dissociate CSCs with trypsin into a single-cell suspension by incubating at 37 ° C for at least 3 min until the cell becomes round.
3. Stop trypsinization by adding a pre-warmed medium containing 10% FBS.
4. Centrifuge cells at 200 × g for 5 min at 25 °C.
5. Remove the supernatant without disturbing the pellet.
6. Resuspend the cells in 1 mL pre-warmed sphere medium.
7. Determine the cell number using hemocytometer.
8. Prepare serially diluted cells in tubes containing the following final cell concentration per tube.

   A. Tube 1 (for 20 cells/100 μL/well ×16 wells).
   B. Tube 2 (for 10 cells/100 μL/well ×16 wells).
   C. Tube 3 (for 5 cells/100 μL/well ×16 wells).
   D. Tube 4 (for 2.5 cell/100 μL/well ×16 wells).

9. Seed the cells into a 96-well ultra-low attachment plate, in the presence of sphere media.
10. Incubate in a 37 °C, 5% CO_2 humidified incubator for 7–10 days.
11. Monitor each well for signs of spheroid formation every day.
12. Count the number of all wells that contain no sphere or sphere greater than ~50 μm diameter at day 10.
13. Take pictures for developed spheres at different cell numbers (Fig. 8.3).
14. Input the number of wells with spheres into ELDA software at http://bioinf.wehi.edu.au/software/elda/ (Fig. 8.3).

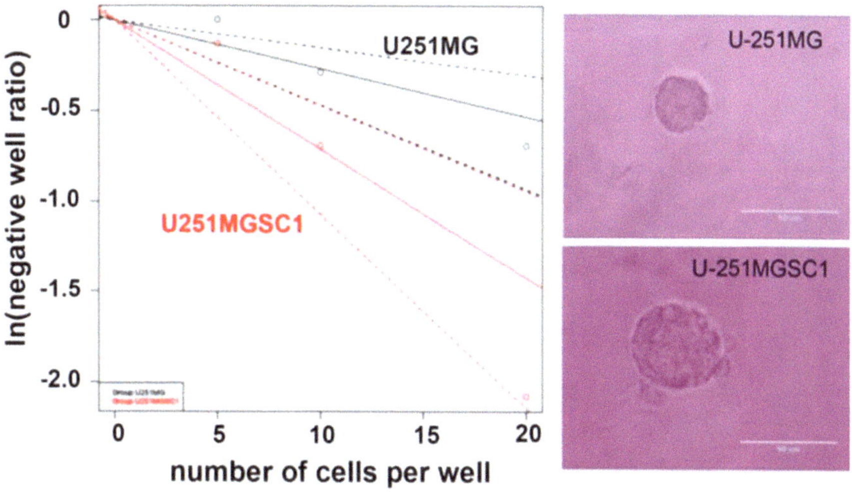

Fig. 8.3 U251 MGSC1 cells are more capable of forming spheres than U251MG cells. To evaluate the ability, a limiting dilution analysis was conducted. Cells were seeded into each 16-well at a rate of 20, 10, 5, and 2.5 cells per well. A 95% confidence interval is shown by the dotted lines. As shown on the right, typical spheres were observed for U-251 MG and U-251 MGSC1 cells. In general, the smaller the value obtained by dividing confidence intervals for 1 by stem cell frequency, the greater the likelihood of sphere formation. The figure is copied from our original paper: Ishii H, Mimura Y, Zahra MH, Katayama S, Hassan G, Afify SM, Seno M. Isolation and characterization of cancer stem cells derived from human glioblastoma. Am J Cancer Res. 2021 Feb 1;11(2):441–457

8.3.4 *Evaluation of Tumorigenic Potential of the Isolated Cells*

Tumorigenic potential is one of the most important characters of cancer stem cells. To assess the potential of the isolated cells, xenograft into nude mice is performed with the parent cells as the control.

1. Prepare U-251MG SC1 cells at 70–80% confluent in a 60-mm dish.
2. Aspirate the medium.
3. Wash the cells with sterile PBS.
4. Trypsinize the cells using 0.025% trypsin-EDTA at 37 °C for 5 min.
5. Stop the reaction by adding normal medium.
6. Transfer the cells into a sterile 15-mL conical tube.
7. Centrifuge the tube at 120×g for 5 min at 25 °C.
8. Aspirate the supernatant without disturbing the pellet.
9. Resuspend the pellet in 1 mL of sterile PBS.
10. Centrifuge the tube at 200 ×g for 5 min at 25 °C.
11. Aspirate the supernatant without disturbing the pellet.
12. Suspend the cells in 1 mL of normal medium.

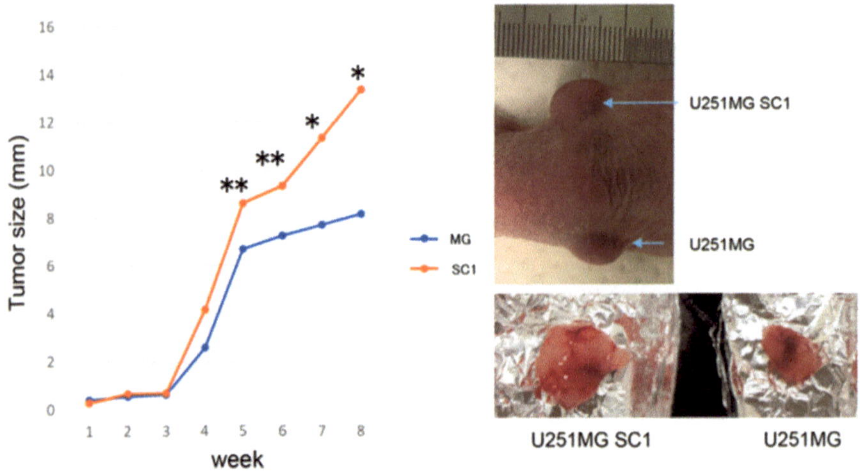

Fig. 8.4 Tumorigenic potential of U-251 MG 1 cells. (**a**) U-251MG cells and U-251MGSC1 cells were subcutaneously transplanted into nude mice and the growth of tumor was monitored by the size from week 1 to 8 after the injection. The tumor size was the average of the vertical and horizontal diameters. $n = 8$. The figure is copied from our original paper: Ishii H, Mimura Y, Zahra MH, Katayama S, Hassan G, Afify SM, Seno M. Isolation and characterization of cancer stem cells derived from human glioblastoma. Am J Cancer Res. 2021 Feb 1;11(2):441–457

13. Take 20 μL of the suspension and count the cells using a hemocytometer.
14. Adjust 2×10^6 cells in 100 μL with sterile HBSS and then transfer the suspension into a 1.5 mL Eppendorf tube.
15. Use parental U-251MG cells as control to compare the tumorigenicity.
16. Keep the tube on ice until the mice get ready for injection (Note 7).
17. Inject the cells subcutaneously.
18. Measure the tumor size every week.
19. Take pictures of the mice with the tumor at any time. Note: Put a measure and when you take pictures of a mouse or a tumor.
20. Excise the tumor for other analysis when it becomes available with suitable volume (Fig. 8.4).

Notes

1. The primary culture is named U-251MG P1 cell.
2. A couple of dishes should be prepared to obtain enough cells to proceed to the next step.
3. The digestion should be as mild as possible so that this step should not impair the cell surface markers, which are necessary for recognition by the antibodies.
4. Two tubes for the cells are separated as staining positive and negative cells.
5. Distinguish the cells from different wells.

6. The scale-up should be performed in the order of 12-well plate, 6-well plate, 60-mm dish, and 100-mm dish according to the cell growth. Skipping the size may delay the growth of cells.
7. The cells should be alive for 1 h on ice.

References

Afify SM, Seno M. Conversion of stem cells to cancer stem cells: undercurrent of cancer initiation. Cancers (Basel). 2019;11(3):345.

Afify SM, Hassan G, Osman A, Calle AS, Nawara HM, Zahra MH, El-Ghlban S, Mansour H, Alam MJ, Abu Quora HA, Du J, Seno A, Iwasaki Y, Seno M. Metastasis of cancer stem cells developed in the microenvironment of hepatocellular carcinoma. Bioengineering (Basel). 2019;6(3):73.

Afify SM, Hassan G, Seno A, Seno M. Cancer-inducing niche: the force of chronic inflammation. Br J Cancer. 2022;127(2):193–201.

Arnold CR, Mangesius J, Skvortsova II, Ganswindt U. The role of cancer stem cells in radiation resistance. Front Oncol. 2020;10:164.

Chiba T, Kita K, Zheng YW, Yokosuka O, Saisho H, Iwama A, et al. Side population purified from hepatocellular carcinoma cells harbors cancer stem cell-like properties. Hepatology. 2006;44(1): 240–51.

Haraguchi N, Utsunomiya T, Inoue H, Tanaka F, Mimori K, Barnard GF, et al. Characterization of a side population of cancer cells from human gastrointestinal system. Stem Cells. 2006;24(3): 506–13.

Ho MM, Ng AV, Lam S, Hung JY. Side population in human lung cancer cell lines and tumors is enriched with stem-like cancer cells. Cancer Res. 2007;67(10):4827–33.

Kondo T, Setoguchi T, Taga T. Persistence of a small subpopulation of cancer stem-like cells in the C6 glioma cell line. Proc Natl Acad Sci U S A. 2004;101(3):781–6.

Lapidot T, Sirard C, Vormoor J, Murdoch B, Hoang T, Caceres-Cortes J, et al. A cell initiating human acute myeloid leukaemia after transplantation into SCID mice. Nature. 1994;367(6464): 645–8.

Mitsutake N, Iwao A, Nagai K, Namba H, Ohtsuru A, Saenko V, et al. Characterization of side population in thyroid cancer cell lines: cancer stem-like cells are enriched partly but not exclusively. Endocrinology. 2007;148(4):1797–803.

Moti N, Malcolm T, Hamoudi R, Mian S, Garland G, Hook CE, et al. Anaplastic large cell lymphoma-propagating cells are detectable by side population analysis and possess an expression profile reflective of a primitive origin. Oncogene. 2014.

Nakatsugawa M, Takahashi A, Hirohashi Y, Torigoe T, Inoda S, Murase M, et al. SOX2 is overexpressed in stem-like cells of human lung adenocarcinoma and augments the tumorigenicity. Lab Investig. 2011;91(12):1796–804.

Nawara HM, Afify S, Hassan G, Zahra MH, Atallah MN, Mansour H, Abu Quora HA, Alam MJ, Osman A, Kakuta H, Hamada H, Seno A, Seno M. Paclitaxel and Sorafenib: the effective combination of suppressing the self-renewal of cancer stem cells. Cancers (Basel). 2020;12(6): 1360.

Nawara HM, Afify SM, Hassan G, Zahra MH, Seno A, Seno M. Paclitaxel-based chemotherapy targeting cancer stem cells from mono- to combination therapy. Biomedicine. 2021;9(5):500.

Nishizawa S, Hirohashi Y, Torigoe T, Takahashi A, Tamura Y, Mori T, et al. HSP DNAJB8 controls tumor-initiating ability in renal cancer stem-like cells. Cancer Res. 2012;72(11): 2844–54.

Phi LTH, Sari IN, Yang YG, Lee SH, Jun N, Kim KS, Lee YK, Kwon HY. Cancer stem cells (CSCs) in drug resistance and their therapeutic implications in cancer treatment. Stem Cells Int. 2018;(2018):5416923.

Sung H, Ferlay J, Siegel RL, Laversanne M, Soerjomataram I, Jemal A, Bray F. Global cancer statistics 2020: GLOBOCAN estimates of incidence and mortality worldwide for 36 cancers in 185 countries. CA Cancer J Clin. 2021;71(3):209–49.

Szotek PP, Pieretti-Vanmarcke R, Masiakos PT, Dinulescu DM, Connolly D, Foster R, et al. Ovarian cancer side population defines cells with stem cell-like characteristics and Mullerian inhibiting substance responsiveness. Proc Natl Acad Sci U S A. 2006;103(30):11154–9.

Yasuda K, Torigoe T, Morita R, Kuroda T, Takahashi A, Matsuzaki J, et al. Ovarian cancer stem cells are enriched in side population and aldehyde dehydrogenase bright overlapping population. PLoS One. 2013;8(8):e68187.

Zahra MH, Nawara HM, Hassan G, Afify SM, Seno A, Seno M. Cancer stem cells contribute to drug resistance in multiple different ways. Adv Exp Med Biol. 2022;1393:125–39.

Chapter 9
Artificial Generation of Cancer Stem Cells from Human Stem Cells

Abstract Due to the presence of cancer stem cells (CSCs), which are considered potent to differentiate into various phenotypes, tumors are considered to be heterogeneous. CSCs are typically described by the self-renewal allowing continuous proliferation, and tumorigenicity as well as differentiation potential. Stem cells are thought to differentiate into tissue-specific phenotype mature cells being affected by the surrounding microenvironment or "niche" surrounding them. Among the factors that control stem cell behavior are neighboring cells, extracellular matrix, as well as biochemical signals. The microenvironment can affect the fate of the stem cell, directing it to become a specific cell type depending on the signals it receives. In this insight, CSCs should be originated from the stem cells under the effect of a chronic inflammatory microenvironment, which stimulates tumor initiation. In this chapter, we will describe this microenvironment as a "cancer inducing niche" and develop a protocol for generating CSCs from stem cells.

In line, we detail the process with critical points in the conversion of stem cell into CSC.

Keywords Human pluripotent stem cells · Cancer · Cancer stem cells

9.1 Introduction

Cancer is the most serious disease with the major cause of mortality worldwide. Since more than 20 million new cases of cancer per year have been predicted by 2030 (Bray et al. 2012), precise therapeutic strategies as well as prevention should urgently be desired to be established.

Nowadays cancer phenotype varies even in the same tissue depending on the genetic background. From the cancer initiation to the development of cancer, the mechanism should be investigated to understand the disease more precisely to find the way leading to personalized medicine. However, the materials to investigate cancer initiation are extremely limited. Cancer initiation should be closely related with the appearance of cancer stem cells (CSCs) (Afify and Seno 2019; Nguyen et al. 2012). CSCs are defined to be responsible for tumorgenicity (Afify et al. 2019; Mansour et al. 2020; Shiozawa et al. 2013) as well as metastasis and heterogeneity

S. M. Afify, M. Seno, *Methods in Cancer Stem Cell Biology*,
https://doi.org/10.1007/978-981-99-1331-2_9

providing different components of microenvironments (Osman et al. 2020a; Yan et al. 2014; Hassan et al. 2019). In this context, CSCs could be the appropriate materials for the investigation if they are available in a simple way.

Normal cells/stem cells are accepted to convert into a CSC originating a tumor (Afify and Seno 2019). And CSCs were the origin of cancer with the abilities to self-renew and differentiate into different phenotypes associated with cancer forming the tumor microenvironment (Osman et al. 2020b).

As the trigger of the induction of CSCs, several issues are hypothesized as the events, such as genomic instability, inflammatory microenvironment, cell fusion, and lateral gene transfer in either stem and/or differentiated cells. Chronic inflammation sounds highly possible reason for the conversion of normal stem cells into CSC because there are many chronic diseases related to initiation of cancer at the end.

The inflammatory microenvironment has successfully been demonstrated as the factor to convert stem cells into CSCs without transducing foreign genes and/or mutations (Chen et al. 2012; Yan et al. 2014; Calle et al. 2016; Nair et al. 2017; Afify et al. 2020, 2022; Minematsu et al. 2022). The CSCs established from stem cells including iPSCs have been defined by the capacity of self-renewal, differentiation, and malignant tumorigenicity. In the current chapter, we summarize the detailed procedure and the inflammatory conditions converting stem cells into CSCs step by step.

9.2 Materials

9.2.1 Reagents

- A-172 (ATCC® CRL-1620™).
- U-87 MG (ATCC® HTB-14™).
- MDA-MB-415(ATCC® HTB-128™).
- MDA-MB-231(ATCC® CRM-HTB-26™).
- T-47D (ATCC® HTB-133™).
- ZR-75-1 (ATCC® CRL-1500™).
- SK-BR-3 (ATCC® HTB-30).
- BT-549 (ATCC® HTB-122).
- MCF7 (ATCC® HTB-22).
- HT-29 (ATCC® HTB-38).
- PANC-1 (ATCC® CRL-1469).
- SK-OV-3 (ATCC® HTB-77).
- Hep G2 (ATCC® HB-8065).
- PLC/PRF/5 (ATCC® CRL-8024).
- Huh-7 cell line (RCB1366, Riken, Japan).
- ECC4 (RCB0982, Riken, Japan).
- CW-2 (RCB0778, Riken, Japan).
- PMF-ko14 (RCB1426, Riken, Japan).

- MY (RCB1701, Riken, Japan).
- MOLT-4 (ATCC® CRL-1582).
- KLM-1 (RCB2138, Riken, Japan).
- PK-8 (RCB2700, Riken, Japan).
- PK-59 (RCB1901, Riken, Japan).
- A549 (ATCC® CRM-CCL-185).
- Li-7 (RCB1941, Riken, Japan).
- NIH:OVCAR-3 (ATCC® HTB-161™).
- Lu99B (RCB1971, Riken, Japan).
- RERF-LC-KJ (RCB1313, Riken, Japan).
- OVK18 (RCB1903, Riken, Japan).
- RERF-LC-AI (RCB0444, Riken, Japan).
- mTeSR™1 medium (STEM-CELL Technologies).
- hiPSCs (409B2, Riken Cell Bank, Tokyo, Japan).
- Mitomycin C treated mouse embryonic fibroblast (MEF) cells (REPROCELL Inc., Kanagawa, Japan).
- Basic FGF (Chemicon, CA, USA).
- Corning® Matrigel®, growth factor reduced, phenol red-free (BD, cat. no. 356231).
- Dulbecco's modified Eagle's medium-high glucose (Wako, Osaka, Japan (catalog number: 044-29765)).
- Trypsin-EDTA (0.25%) (Nacalai Tesque, Kyoto, Japan, Cat. No: 327777-44).
- Fetal bovine serum (FBS, Gibco, Life Technologies, Massachusetts, USA, Cat. No: 10437-028).
- Penicillin/streptomycin mixed solution (100 U/mL) Nacalai Tesque, Kyoto, Japan, Cat. No-26253-84).
- 70% ethanol (Sigma-Aldrich; Cat. No.: 459836-2).
- Liquid N_2 storage tank.
- Hank's balanced salt solution (HBSS) Genesee Scientific, El Cajon, USA.

9.2.2 Equipment

- Eppendorf Centrifuge 5415R, Eppendorf AG, 22331 Hamburg, Germany.
- Sanyo MCO-19AIC(UV) CO2 Incubator, Marshall Scientific, Hampton, USA.
- Type A2 Biological Safety Cabinets (E-Series).
- Olympus IX81 microscope (Olympus, Tokyo, Japan).
- Tissue culture-treated plate, 60 mm dish; TPP, AG Schaffhausen, Switzerland, Cat. No 93060.
- Tissue culture-treated plate, 100 mm dish; TPP, AG Schaffhausen, Switzerland, Cat. No 93100.
- Filter max 250 mL, TPP, Switzerland, Cat. No 99255.
- Falcon® Conical Centrifuge Tubes (15 mL; BD Falcon, New York, USA, Cat. No 352095).

- Falcon® Conical Centrifuge Tubes (50 mL; BD Falcon, New York, USA, Cat. No 352070).
- 37 °C water bath.

9.2.3 *Reagent Preparation*

MEF Medium (DMEM + 10% FCS)
Mix the following

DMEM (Sigma: #D5796)	445 mL
FBS	50 mL
Penicillin/streptomycin (Life Technologies)	5 mL

Repro Stem Cell Medium
Mix the following for 500 mL

Repro stem cell medium (#RCHEMD001)	495 mL
BFGF, 0.5 mg/ml (final 5 ng/ml)	5 mL

StemFit Medium
Mix the following for 500 mL

StemFit AK02N Basal (RCAK02N)	495 mL
BFGF, 0.5 μg/ml (final 5 ng/ml)	5 mL

Dissociation Solution Mix the following for 100 mL

Trypsin (Difco; #215240)	0.25 g
Collagenase IV, 10 mg/ml (Life Technologies, #17104–019)	10 mL
KnockOut Serum Replacement (Life Technologies)	20 mL
1 M $CaCl_2$ (Final 1 mM)	0.1 mL
PBS	70 mL

9.3 Methods

This protocol describes how to generate CSCs from human iPSCs in the presence of microenvironments induced by different cancer cells. To convert stem cells into CSCs, the protocol outlines detailed experimental steps. The first step was to collect CM from a confluent culture of cancer cells. CSCs were then generated from human iPSCs by maintaining them with CM (Fig. 9.1).

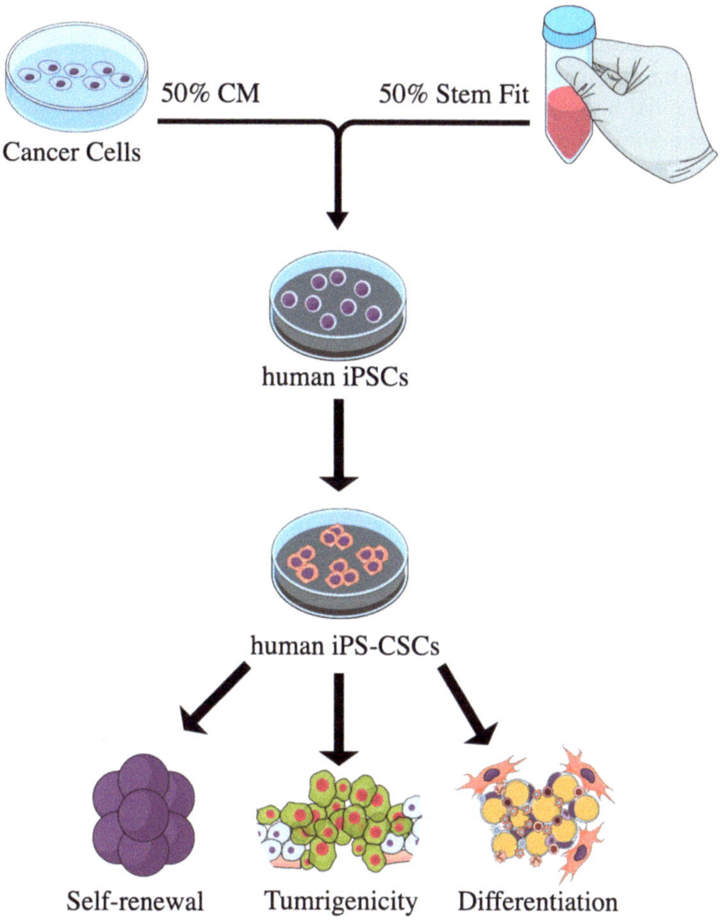

Fig. 9.1 Diagram showing the preparation of human CSCs from human iPS cells, along with their potential

9.3.1 Preparation of Gelatin-Coated Dishes

1. Transfer 0.2% gelatin solution from 4 °C to the clean bench.
2. Cover the bottom of the dish with sufficient volume of 0.2% gelatin solution (Note 1).
3. Incubate the dishes for 30 min at 37 °C in a 5% CO_2 incubator.

9.3.2 Preparation of MEF-Coated Dishes

1. Pick up a storage vial of mitotically inactivated MEF from liquid nitrogen and put on ice.

2. Immerse the vial quickly in a 37 °C water bath without submerging the cap.
3. Remove the vial from the water bath just before the ice is completely dissolved.
4. Detox the surface of the vial with 70% ethanol.
5. Put the vial into the laminar hood.
6. Prepare 5 mL of pre-warmed MEF medium in a sterile 15-mL conical tube.
7. Pipette up the thawed cells gently and transfer into the conical tube containing MEF medium (Note 2).
8. Mix the cells by slow and gentle pipetting in the MEF medium.
9. Centrifuge the cells at 200 ×g for 5 min at 25 °C.
10. Aspirate the supernatant.
11. Resuspend the cell pellet with 2 mL of fresh MEF medium.
12. Count the cells using hemocytometer.
13. Pick up the gelatin-coated 6-well plate from the incubator.
14. Aspirate the gelatin solution.
15. Seed MEF at 1×10^6 cells/well.
16. Add 3 mL of MEF medium per well.
17. Incubate the plates at 37 °C in a 5% CO_2 incubator.
18. Use the plates with MEF feeder layer within 3–4 days after seeding (Note 3).

9.3.3 Reviving Human iPSCs

1. Pre-warm stem cell medium at 37 °C in a water bath for 30 min (Note 4).
2. Pick up the vial of human iPSCs from liquid nitrogen storage.
3. Thaw the cells quickly at 37 °C in a water bath.
4. Sterilize the vial with 70% ethanol.
5. Add 5 mL of pre-warmed stem cell medium to a sterile 15-mL conical tube.
6. Suspend the cells gently with a sterile long Pasteur pipette.
7. Transfer the cells into the 15-mL conical tube containing the medium.
8. Centrifuge cells at 200 ×g for 5 min at 25 °C.
9. Aspirate the supernatant without disturbing the pellet.
10. Add 5 mL of pre-warmed stem cell medium.
11. Resuspend the pellet by gentle pipetting up and down 2 or 3 times using a sterile long Pasteur pipette.
12. Pick up the MEF-coated dishes prepared in the previous step from the incubator.
13. Replace the MEF medium with 4 mL of Repro stem medium containing 5 ng/mL of human FGF2 and 10 μM of rock inhibitor Y-27632.
14. Seed the cells on two MEF-dishes with high and low numbers. Note: Avoid seeding human iPSCs at a high density because they tend to aggregate and give rise to cells with heterogeneous morphologies.
15. Remove the dead cells by the medium change on the next day.
16. Monitor the cells until the colonies become visible and bright.

9.3.4 Passage of Human iPSCs on MEF

1. Passage human iPSCs when the colonies become large and bright (Note 5).
2. Take the dish out from the incubator.
3. Aspirate the medium.
4. Wash the dish twice with 2 mL of PBS.
5. Aspirate PBS.
6. Add 250 μL of dissociation buffer/60-mm dish.
7. Incubate the dish for 5 min in the incubator at 37 °C under 5% CO_2.
8. Take the dishes out from the incubator.
9. Aspirate the dissociation buffer.
10. Wash the colonies of human iPSCs with PBS.
11. Add 1 mL of pre-warmed stem culture medium.
12. Detach the colonies of human iPSCs with a cell scraper.
13. Collect the colonies and make them small by gentle pipetting.
14. Take MEF-coated dishes out from the incubator.
15. Aspirate MEF medium.
16. Add 4 mL Repro Stem medium.
17. Divide the small colonies into MEF-coated dishes.
18. Add FGF2 to 5 ng/mL and rock inhibitor, Y-27632, to 10 μM into every 5-mL Repro Stem medium.
19. Incubate the dishes in the incubator at 37 °C under 5% CO_2.
20. Change the medium to Repro Stem medium without Y-27632 on the next day.

9.3.5 Passage of Human iPSCs on Matrigel

1. Pick up Matrigel from a freezer and place at 4 °C 1 day before transferring human iPSCs.
2. Add 2 ml of Repro Stem medium into a 15-ml tube placed on ice.
3. Add 50 ul Matrigel with cold tips.
4. Add 2 ml Repro Stem medium into a 60-mm dish containing Matrigel.
5. Incubate the dish in an incubator at 37 °C with 5% CO_2 for at least 1 h.
6. Take the dish of human iPSCs out from the incubator.
7. Aspirate the medium from the dish.
8. Wash the dish twice with PBS.
9. Add 250 μL of dissociation buffer to a 60-mm dish.
10. Spread the buffer well over the surface.
11. Incubate the dish in the incubator at 37 °C with 5% CO_2 for 2–3 min until MEF cells detach.
12. Aspirate the dissociation buffer.
13. Wash the colonies of human iPSCs with PBS.
14. Detach the colonies of human iPSCs with a cell scraper.
15. Collect the colonies and make them small by gentle pipetting.

16. Take the Matrigel-coated dish out from the incubator.
17. Aspirate media from the dish.
18. Wash the dish with PBS.
19. Add 4 ml of StemFit media into the dish.
20. Divide human iPSC colonies into three Matrigel-coated dishes.
21. Add FGF2 to 5 ng/mL and rock inhibitor, Y-27632, to 10 μM into every 5-mL StemFit medium.
22. Incubate the dish in an incubator at 37 °C with 5% CO_2.
23. Change medium to StemFit medium without Rock inhibitor Y-27632.

9.3.6 Preparation of Conditioned Medium (CM) for Conversion

1. Take a vial of each cancer cell line from liquid nitrogen storage.
2. Revive the cells into a 60-mm dish containing 5 mL Dulbecco's modified Eagle's medium (DMEM) supplemented with 10% FBS and 100 U/mL penicillin/streptomycin.
3. Passage cells into a 100-mm dish after a couple of days, when the cells become 70% confluent.
4. Change the medium to 5% FBS at 80% confluency prior to the collection of conditioned mediums.
5. Collect the conditioned medium (CM) from the confluent dishes after 48-h incubation.
6. Centrifuge the CM at 300 g for 10 min at 25 °C.
7. Filter the supernatant through a 0.22 mm filter.
8. Add the filtered CM to a 35-mm dish and incubate overnight in an incubator at 37 °C with 5% CO_2 to confirm no surviving cells in the CM.
9. Store the CM at −20 °C until use.

9.3.7 Conversion of Human iPSCs on Matrigel

1. Start conversion of human iPSCs in the presence of CM when the colonies of human iPSCs become 70% confluent.
2. Aspirate the medium from the dish.
3. Wash the dish twice with PBS.
4. Add 250 μL of dissociation buffer to a 60-mm dish.
5. Spread the buffer well over the surface.
6. Incubate the dish in an incubator at 37 °C with 5% CO_2 for 2–3 min until MEF cells detach.

7. Aspirate the dissociation buffer.
8. Wash the colonies of human iPSCs with PBS.
9. Detach the colonies of human iPSCs with a cell scraper.
10. Collect the colonies and make them small by gentle pipetting.
11. Take the Matrigel-coated dish out from the incubator.
12. Aspirate media from the dish.
13. Wash the dish with PBS.
14. Add 2 mL of StemFit and 2 mL of CM from cancer cell culture to each Matrigel-coated dish.
15. Divide human iPS colonies into three Matrigel-coated dishes.

Observe the cells every day while taking pictures (Table 9.1) (Note 6).

This colony-forming population was found to contain Podocalyxin, a glycoprotein that is recognized by the BC2LCN-635 lectin as a marker of human stem cells. CSCs expressing Podocalyxin could be manually selected in order to isolate them (Fig. 9.2).

Cells should be assessed for the self-renewal potential by sphere formation assay. Human CSCs should show the self-renewal potential by forming tumor spheroids within 7 days from culture in suspension culture (Fig. 9.3).

In order to conclude that the cells derived from human iPSCs are CSCs, further evaluation of tumorigenicity in vivo is necessary. In case these convert to CSCS, malignant tumors should be detected after 2 months (Figs. 9.4 and 9.5).

Furthermore, human CSC could be detected in the tumor tissue with rBC2LCN in the tumor derived from OCC-hiPS-1 injection in testis when compared to normal testis as shown in Fig. 9.6.

Notes

1. For example, 2 mL for a 60-mm dish and 5 mL for a 100-mm dish.
2. This step should be done as quickly as possible so as not to get the cells stressed.
3. Plates with MEF feeder layer can be kept up to approx. 2 weeks in the incubator until use.
4. Do not keep the media in the water bath for more than 1 h at 37 °C as continued exposure to 37 °C may impair the components and reduce the activity of the growth factors.
5. Do not wait until the colonies become dark because the cells will start to die.
6. In the microenvironment they were grown in, human iPS cells may have exhibited different phenotypes.

Table 9.1 This table illustrates human iPSCs and related human CSCs generated in the presence of various cancer-derived conditioned media

CSCs	Cell line	Organ	CM	BF for CSCs	Number per injection
OCC-hiPS-1	A172	Glioma	CM		1.0×10^4 (Balb/c-nu/nu) 1.0×10^4 (CB.17 SCID)
OCC-hiPS-2	U251-MG	Glioma	CM		5.5×10^4 (Balb/c-nu/nu) 1.0×10^4 (CB.17 SCID)
OCC-hiPS-3	MDA-MB-415	Breast	CM		1.0×10^4 (NOD SCID)
OCC-hiPS-4	MDA-MB-231	Breast	CM		1.0×10^4 (NOD SCID)
OCC-hiPS-5	T-47D	Breast	CM		1.0×10^4 (CB.17 SCID)
OCC-hiPS-6	ZR-75-1	Breast	CM		3.5×10^5 (CB.17 SCID)
OCC-hiPS-7	SK-BR-3	Breast	CM		3.3×10^5 (Balb/c-nu/nu)
OCC-hiPS-8	BT-549	Breast	CM		1.0×10^4 (NOD SCID)
OCC-hiPS-9	MCF7	Breast	CM		6.7×10^5 (Balb/c-nu/nu)
OCC-hiPS-10	HT-29	Colon	CM		1.0×10^5, 8.0×10^4 (Balb/c-nu/nu) 8.8×10^5, 7.7×10^4 (CB.17 SCID)
OCC-hiPS-11	PANC1	Pancreas	CM		2.5×10^3 (Balb/c-nu/nu)
OCC-hiPS-12	SKOV3	Ovary	CM		7.9×10^5, 2.7×10^5 (Balb/c-nu/nu) 1.1×10^5 (CB.17 SCID) 1.0×10^4 (NOD SCID)
OCC-hiPS-13	HepG2	hepatocellular	CM		2.0×10^3 (Balb/c-nu/nu)
OCC-hiPS-14	Huh-7	hepatocellular	CM		2.5×10^4 (Balb/c-nu/nu)
OCC-hiPS-15	PLC/PRF-5	hepatocellular	CM		5.0×10^3 (Balb/c-nu/nu)

(continued)

Table 9.1 (continued)

CSCs	Cell line	Organ	CM	BF for CSCs	Number per injection
OCC-hiPS-16	ECC4	Gastrointestinal	CM		1.0×10^4 (NOD SCID)
OCC-hiPS-17	CW-2	Colon	CM		1.0×10^4 (NOD SCID)
OCC-hiPS-18	PMF-ko14	Spleen	CM		-
OCC-hiPS-19	MY	Lymphocyte	CM		1.0×10^4 (NOD SCID)
OCC-hiPS-20	MOLT4	T-cell leukemia	CM		1.0×10^4 (NOD SCID)
OCC-hiPS-21	KLM-1	Pancreas	CM		-
OCC-hiPS-22	PK8	Pancreas	CM		-
OCC-hiPS-23	PK59	Pancreas	CM		1.0×10^4 (NOD SCID)
OCC-hiPS-24	A549	Lung	CM		-
OCC-hiPS-25	Li-7	Hepatocellular	CM		1.0×10^4 (NOD SCID)
OCC-hiPS-26	NIH: OVCAR-3	Ovary	CM		-
OCC-hiPS-27	Lu99B	Lung	CM		-
OCC-hiPS-28	RERF-LC-KJ	Ovary	CM		-
OCC-hiPS-29	A172	Glioma	CM/primary culture (CB.17 SCID)		-
OCC-hiPS-30	OVK18	Ovary	CM		-
OCC-hiPS-31	RERF-LC-AI	Lung	CM		-

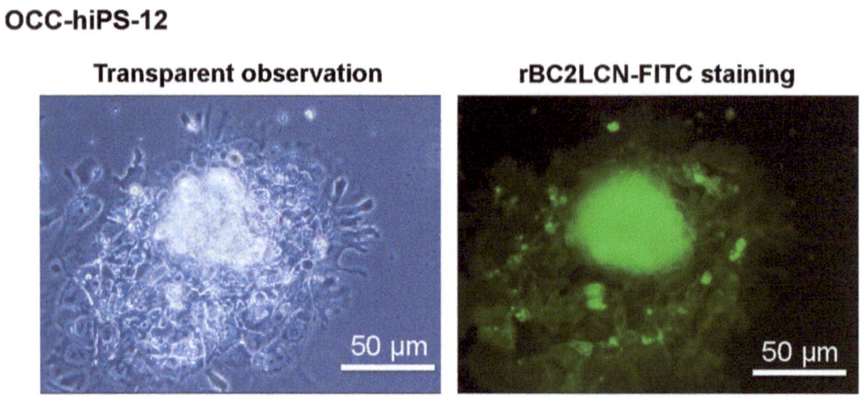

Fig. 9.2 Representative image of human iPSCs cultured in the presence of CM from ovary cancer cell line SKOV3 cells. Podocalyxin is a sialoglycoprotein specific to pluripotent stem cells. BC2LCN is a lectin specific to Podocalyxin (Onuma et al. 2013, Biochem Biophys Res Commun. 431(3):524–529)

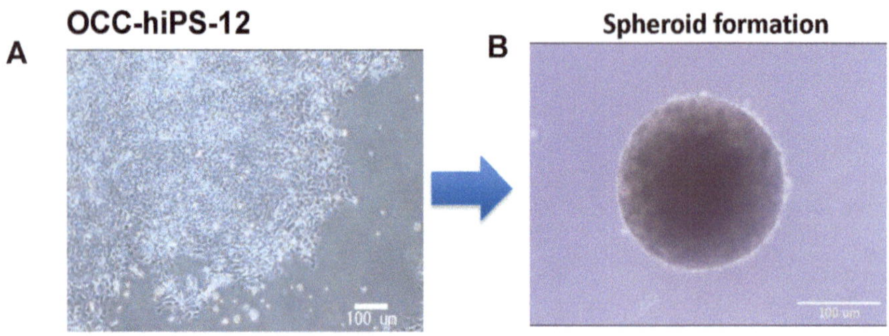

Fig. 9.3 (**a**) Representative image of human iPSCs in the presence of CM from ovary carcinoma cell culture. (**b**) Representative image of sphere derived from CSCs generated from human iPSCs in the presence of CM derived from ovary carcinoma cell line SKOV3 cell culture. OCC: *O*kayama *C*ancer stem cell *C*ollection

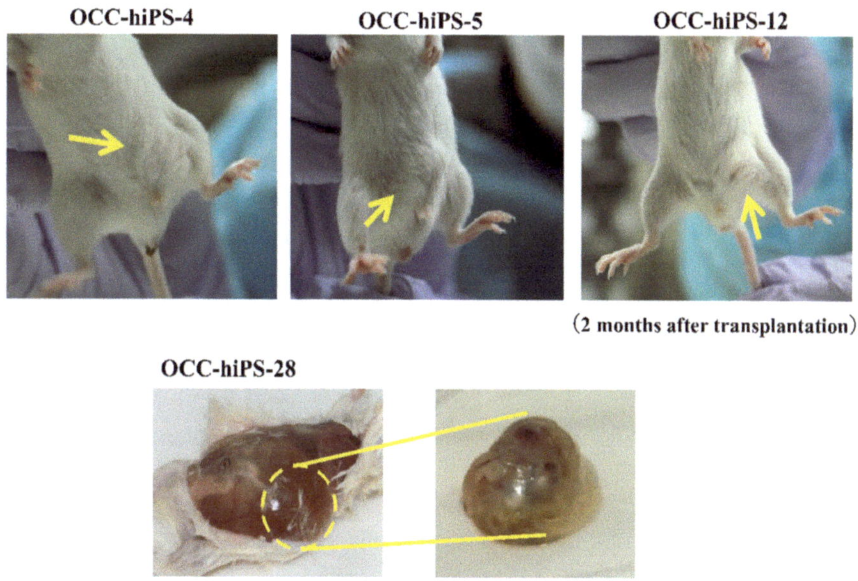

Fig. 9.4 Represented image for NOD SCID mouse injected by human CSCs (OCC-hiPS-4, OCC-hiPS-5, OCC-hiPS-12, OCC-hiPS-28) generated in the presence of breast and/or ovary carcinoma

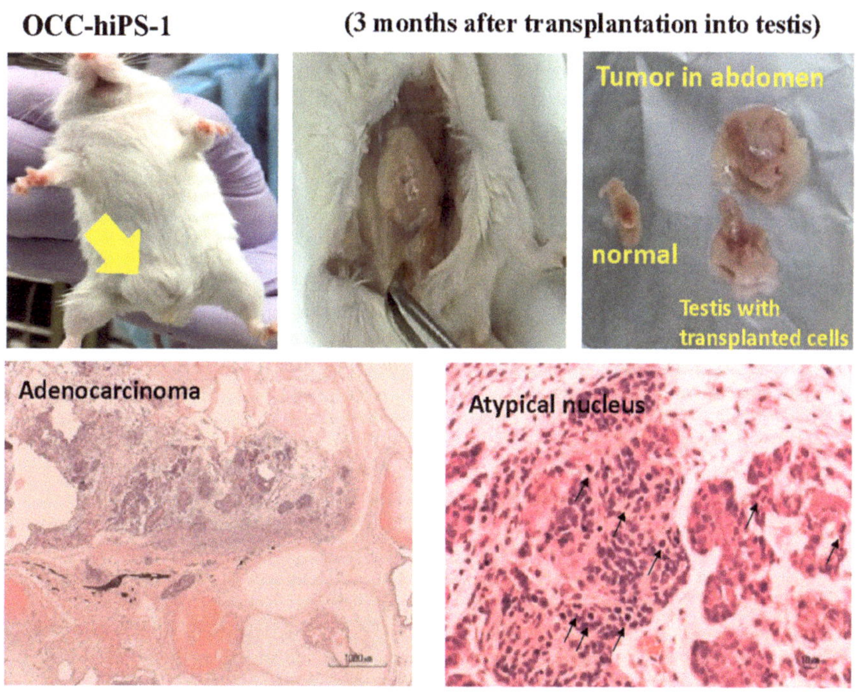

Fig. 9.5 (**a**) represented image for CB.17 SCID mouse injected by human CSCs (OCC-hiPS-1) generated in the presence of Glioma CM. (**b**) Histological evaluation of the malignant tumor derived from human CSC (OCC-hiPS-1)

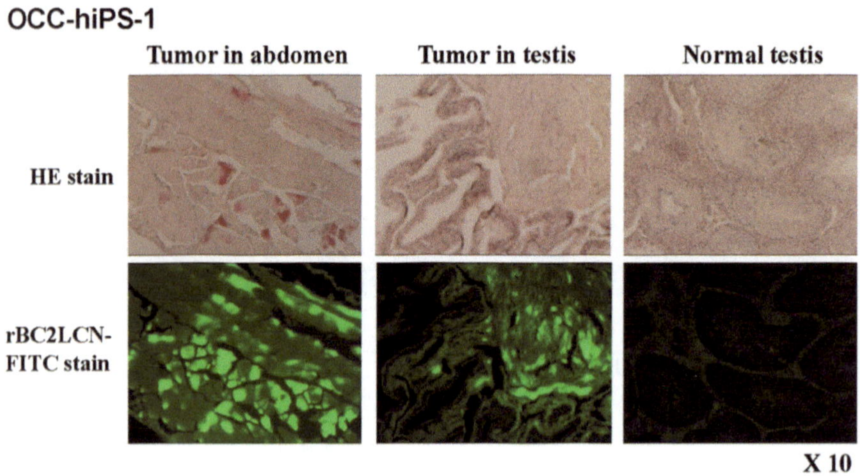

Fig. 9.6 (**a**) Representatives images for H&E of tumors derived from OCC-hiPS-1 in testis and abdomen compared to normal testis. (**b**) Immunofluorescence staining for OCC-hiPS-1 derived tumor tissue with BC2LCN

References

Afify SM, Seno M. Conversion of stem cells to cancer stem cells: undercurrent of cancer initiation. Cancers (Basel). 2019;11(3):345.

Afify SM, Chen L, Yan T, Calle AS, Nair N, Murakami C, Zahra MH, Okada N, Iwasaki Y, Seno A, Seno M. Method to convert stem cells into cancer stem cells. Methods Protoc. 2019;2(3):71.

Afify SM, Sanchez Calle A, Hassan G, Kumon K, Nawara HM, Zahra MH, Seno M. A novel model of liver cancer stem cells developed from induced pluripotent stem cells. Br J Cancer. 2020;122 (9):1378–90.

Afify SM, Hassan G, Seno A, et al. Cancer-inducing niche: the force of chronic inflammation. Br J Cancer. 2022;127:193–201.

Bray F, Jemal A, Grey N, Ferlay J, Forman D. Global cancer transitions according to the human development index (2008–2030): a population-based study. Lancet Oncol. 2012;13:790–801.

Calle AS, Nair N, Oo AK, Prieto-Vila M, Koga M, Khayrani AC, Hussein M, Hurley L, Vaidyanath A, Seno A, et al. A new PDAC mouse model originated from iPSCs-converted pancreatic cancer stem cells (CSCcm). Am J Cancer Res. 2016;6:2799–815.

Chen L, Kasai T, Li Y, Sugii Y, Jin G, Okada M, Vaidyanath A, Mizutani A, Satoh A, Kudoh T, et al. A model of cancer stem cells derived from mouse induced pluripotent stem cells. PLoS One. 2012;7:e33544.

Hassan G, Afify SM, Nair N, Kumon K, Osman A, Du J, Mansour H, Abu Quora HA, Nawara HM, Satoh A, Zahra MH, Okada N, Seno A, Seno M. Hematopoietic cells derived from cancer stem cells generated from mouse induced pluripotent stem cells. Cancers (Basel). 2019;12(1):82.

Mansour H, Hassan G, Afify SM, Yan T, Seno A, Seno M. Metastasis model of cancer stem cell-derived tumors. Methods Protoc. 2020;3(3):60.

Minematsu H, Afify SM, Sugihara Y, et al. Cancer stem cells induced by chronic stimulation with prostaglandin E2 exhibited constitutively activated PI3K axis. Sci Rep. 2022;12:15628.

Nair N, Calle AS, Zahra MH, Prieto-Vila M, Oo AKK, Hurley L, Vaidyanath A, Seno A, Masuda J, Iwasaki Y, Tanaka H, Kasai T, Seno M. A cancer stem cell model as the point of origin of cancer-associated fibroblasts in tumor microenvironment. Sci Rep. 2017;7(1):6838.

Nguyen LV, Vanner R, Dirks P, Eaves CJ. Cancer stem cells: an evolving concept. Nat Rev Cancer. 2012;12:133–43.

Osman A, Afify SM, Hassan G, Fu X, Seno A, Seno M. Revisiting cancer stem cells as the origin of cancer-associated cells in the tumor microenvironment: a hypothetical view from the potential of iPSCs. Cancers (Basel). 2020a;12(4):879.

Osman A, Oze M, Afify SM, Hassan G, El-Ghlban S, Nawara HM, Fu X, Zahra MH, Seno A, Winer I, Salomon DS, Seno M. Tumor-associated macrophages derived from cancer stem cells. Acta Histochem. 2020b;122(8):151628.

Shiozawa Y, Nie B, Pienta KJ, Morgan TM, Taichman RS. Cancer stem cells and their role in metastasis. Pharmacol Ther. 2013;138:285–93.

Yan T, Mizutani A, Chen L, Takaki M, Hiramoto Y, Matsuda S, Seno M. Characterization of cancer stem-like cells derived from mouse induced pluripotent stem cells transformed by tumor-derived extracellular vesicles. J Cancer. 2014;5(7):572–84. https://doi.org/10.7150/jca.8865.

Chapter 10
Quick Method to Assess Non-mutagenic Carcinogens with iPS Cells

Abstract Carcinogenesis refers to a complex multistage process in which normal cells transform into cancer cells. The transformation will happen in a number of cellular, genetic, and epigenetic levels resulting in uncontrolled cell division to form a malignant tumor. In spite of the fact that the process of carcinogenesis has been extensively studied, the mechanisms underlying it, except for the underlying genetic disorders, remain poorly understood. Even many chemicals are approved and clinically used, many patients experience side effects during the treatment with or without effectiveness. Further, in worse cases, the patients may get another cancer if the treatment continues for a long period.

It is not predictable who might get a second cancer and which therapy might induce it. But higher risk for second cancers sometimes follows after cancer treatment. This might be due to the lack of appropriate assessments of chemicals for carcinogenesis. Considering that more new treatments will emerge, and standard treatments continue to be used, continued studies should be necessary to investigate not only genetics but also the chemicals and the interaction of different chemicals without affecting genetics directly.

However, there are few efficient methods available to evaluate the non-mutagenic carcinogenic effect of chemicals while several methods are available to assess mutagens. Furthermore, most of the current methods require a long period to get results, so the development of new quick methods should eagerly be expected to assess the carcinogenic risk of chemicals. The quick methods of short-term assay were proposed and developed in the last few years as alternatives or supplementary to the traditional long-term bioassays. In these methods, after being exposed to chemicals for a fixed length of period, cells are visually scored according to the specific parameters relating to phenotypes and growth patterns. However, many chemicals judged as safe by these methods are unfortunately suspected later to have possible risks of cancer.

Therefore, it is urgently demanded to develop a quick method or technology to assess the carcinogenic potential of chemicals. In this chapter, we will try to introduce the procedure step by step using or testing chemicals exploiting the procedure of converting stem cells into cancer stem cells.

Keywords Carcinogenesis · Short-term · iPS cells · CSCs

S. M. Afify, M. Seno, *Methods in Cancer Stem Cell Biology*,
https://doi.org/10.1007/978-981-99-1331-2_10

10.1 Introduction

The transformation of a normal cell into a cancer cell is considered as a multi-step process that involves initiation, promotion, and progression. Starting with a single cell, this process is believed to take many years. Tumor initiation can be driven by the transformation of either differentiated cells or stem cells within the normal tissue (Hanahan and Weinberg 2011). The transformation can take place during tissue damage and healing initiated as a response to infections, toxications, radiation, or metabolic influences causing mutations of genetic disorders (Basu 2018). During the transformation, oncogenes may be upregulated and/or tumor suppressors may be inactivated resulting in the uncontrolled cell growth (Hanahan and Weinberg 2011). The progress in cancer research introduced the concept of CSCs development from stem cells during chronic inflammatory microenvironment (Afify and seno 2019; Afify et al. 2019, 2022), cell fusion (Wang 2010), and/or mutation (Greaves and Maley 2012; Collisson et al. 2012). CSCs were first identified in acute myeloid leukemia in 1994 (Lapidot et al. 1994). Then in 1997, Bonnet and Dick identified the CSCs by the ability of self-renewal from a heterogeneous tumor xenograft (Bonnet and Dick 1997). Studies have demonstrated that the CSCs exist in various types of cancers, including breast, brain, lung, gastric, and colorectal cancers (Lee et al. 2011).

Identification and evaluation of the process of cancer initiation, which is the CSC development from stem cells, is essential to find the risks and is helpful in drug screening. In the chemical substance database (https://www.cas.org/), the Chemical Abstracts Service (CAS) and the American Chemical Society registered 100 million chemical substances by 2015. Five years later, the number registered in the database became more than 200 million. One new chemical substance sounded registered every 2 min, the pace of which would make the number more than one billion after 10 years. Many of the registered chemical substances are currently produced by industries and provided to the environment surrounding us. Also, this number is increasing so much that the safeties of these chemical compounds should efficiently be confirmed. Some chemical compounds are well known as mutagens. Some as inhibitors of enzymes such as kinases, phosphatase, proteases, nucleases, and ATPases and others as inhibitors of ion channels on the cytoplasmic membrane. Therefore, chemical compounds are usually assessed for their risks of toxicities such as carcinogens, neurotoxins, and endocrine disruptors. This is the reason why we need various methods to evaluate the risks. For the assessment of cancer risk, there are some methods to evaluate mutagens. However, there are a few efficient methods available to evaluate the non-mutagenic carcinogens. Furthermore, it takes a long time to get results by the current methods, so the development of new quick methods should eagerly be expected to assess the carcinogenic risk of chemicals. The quick methods were proposed and developed in the last few years as alternatives or supplementary to the traditional biological assays. In these methods, cells are visually scored according to the specific parameters relating to phenotypes and growth patterns after exposure to chemicals for a fixed length of period. Although many chemicals have been judged safe by these methods, some of them are unfortunately found to induce cancer later.

We have previously been investigating the CSCs with CSC models that we developed from iPSCs and finding new insights in the CSCs. CSCs are generally considered as a small population of cells present in malignant tumors exhibiting self-renewal and differentiation potential responsible for the malignancy. However, the mechanism inducing CSCs is not well understood yet. Considering that chronic inflammation is the imbalanced abnormal microenvironment lasting a long time, the stem cells or progenitors are affected by the microenvironment to regulate the proliferation, self-renewal potential, and differentiation in each tissue. Finally, stem cells or progenitors are plastic enough to change the characters to convert into CSCs.

In this context, the microenvironment distorted by the factors secreted in the inflammation should be responsible for the induction of CSCs. Since in many cases the factors are known to be secreted or produced by cancer cells, the conditioned medium (CM) of cancer cells is feasible to mimic the inflammation and the long period of exposure to the CM could mimic chronic situations. Therefore, stem cells could become to have the characteristics of CSCs after the long exposure of the CM. In this context, the conversion of mouse induced pluripotent stem cells (miPSCs) into CSCs was demonstrated by the treatment of iPSCs with a CM prepared from different cancer cell lines (Chen et al. 2012; Calle et al. 2016; Nair et al. 2017; Afify et al. 2020). As we mentioned before there is an urgent demand to develop a quick method or technology to assess the carcinogenic potential of chemicals. We propose a simple quick method to assess the risks of carcinogens as cancer-initiating factors applying the process of the conversion to evaluate the conversion of miPSCs into CSCs. This method could be very unique to evaluate the accelerating effect of compounds on the cancer initiation.

10.2 Materials

10.2.1 Reagents

- Corning® Matrigel®, growth factor reduced, phenol red-free (BD, cat. no. 356231).
- Dulbecco's modified Eagle's medium-high glucose (Wako, Osaka, Japan (catalogue number: 04429765)).
- Trypsin-EDTA (0.25%) (Nacalai Tesque, Kyoto, Japan, Cat. No: 327777-44).
- Fetal bovine serum (FBS, Gibco, Life Technologies, Massachusetts, USA, Cat. No: 10437-028).
- Penicillin/streptomycin mixed solution (100 U/mL) Nacalai Tesque, Kyoto, Japan, Cat. No-26253-84).
- 70% ethanol (Sigma-Aldrich; Cat. No.: 459836-2).
- Liquid N_2 storage tank.

- Hank's balanced salt solution (HBSS) Genesee Scientific, El Cajon, USA.
- Endothelial basal medium EBM2 media (EBM-2 SingleQuots Kit, Lonza, Switzerland).

10.2.2 Equipment

- Eppendorf Centrifuge 5415R, Eppendorf AG, 22331 Hamburg, Germany.
- Sanyo MCO-19AIC (UV) CO_2 Incubator, Marshall Scientific, Hampton, USA.
- Type A2 Biological Safety Cabinets (E-Series).
- Olympus IX81 microscope (Olympus, Tokyo, Japan).
- Tissue culture-treated plate, 60 mm dish; TPP, AG Schaffhausen, Switzerland, Cat. No 93060.
- Tissue culture-treated plate,100 mm dish; TPP, AG Schaffhausen, Switzerland, Cat. No 93100.
- Filter max 250 mL, TPP, Switzerland, Cat. No 99255.
- Falcon® Conical Centrifuge Tubes (15 mL; BD Falcon, New York, USA, Cat. No 352095).
- Falcon® Conical Centrifuge Tubes (50 mL; BD Falcon, New York, USA, Cat. No 352070).
- 37 °C water bath.

10.2.3 Reagent Preparation

Stem Cell Medium

Mix the following for 500 mL		
Dulbecco's modified Eagle's medium-high glucose		412.5 mL
FBS		75 mL
Penicillin/streptomycin mixed solution	(50 U/mL)	2.5 mL
Gibco™ L-Glutamine (200 mM)	(2 mM)	5 mL
MEM non-essential amino acids solution	(1 mM)	5 mL

10.3 Methods

There are three steps in this quick method (Fig. 10.1). The first step is the development of a carcinogenic process in vitro. The second is the fixation of the criteria. And the last is the quantification of carcinogenicity residing in the target chemicals by criteria in vitro.

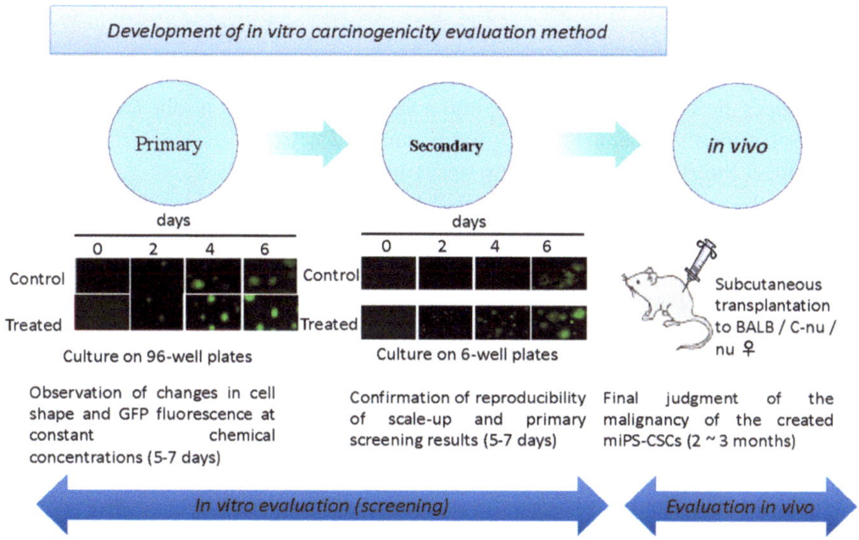

Fig. 10.1 Representative scheme for carcinogen evaluation method

In this method we are using iPSCs to test different chemicals which inhibit the cytoplasmic signaling and are used as anti-cancer agents. iPSCs are generally converted into cancer stem cells (CSCs) in 4 weeks in the presence of CM of cancer-derived cell line Lewis lung carcinoma (LLC) cells. This method evaluates the acceleration of this conversion by the inhibitors in 1 week. The conversion can be confirmed by growing cells without inhibitors in sufficient numbers to demonstrate tumorigenicity in vivo.

10.3.1 Conditioned Medium (CM) Preparation

1. Take a vial of cancer cell line LLC cells from liquid nitrogen storage.
2. Thaw the cells quickly in a 37 °C water bath.
3. Remove the vial from the water bath.
4. Sterilize the tube with 70% ethanol.
5. Transfer the cells with 5 mL of DMEM medium containing 10% FBS to a 15 mL conical tube with a sterilized long Pasteur pipette.
6. Centrifugate the conical tube at 100 *g* for 5 min at 25 °C.
7. Discard the supernatant without disturbing the pellet.
8. Suspend the cells in 5 mL of DMEM containing 10% FBS.
9. Seed the cells into two 60-mm dishes dividing the cells into the ratio approximately 1:2 so that you can prepare a dish with sufficient growth of the cells.
10. Passage the cells after a couple of days at 80% confluency.
11. Repeat the passage two times before starting to collect the CM.

12. Change the medium to 5% FBS at 80% confluency prior to collecting the CM (Note 1).
13. Incubate the cells for 48 h at 37 °C in a 5% CO2 incubator.
14. Collect the CM from LLC cells into a sterile 50-mL Falcon tube (Note 2).
15. Centrifuge the tube at 15,000 g for 10 min at 25 °C.
16. Collect the supernatant in a new sterile tube.
17. Pass the supernatant through a sterile 0.22 μm syringe-top filter to remove cell debris.
18. Pick up 2 mL of CM, transfer to a new 3.5 cm dish and incubate overnight to confirm there are no surviving cancer cells.
19. Store CM at −20 °C until use.

10.3.2 Preparation of Dishes with Feeder Cells

1. Add 2 mL of sterile 0.1% gelatin per 60-mm dish (Note 3).
2. Incubate the dishes for at least 30 min at 37 °C to coat the dishes with gelatin.
3. Take a vial of mitomycin C-treated mouse embryonic fibroblasts (MEFs) from the liquid nitrogen.
4. Thaw quickly in a 37 °C water bath.
5. Remove the vial from the water bath.
6. Sterilize the tube with 70% ethanol.
7. Transfer the cells to a 15-mL conical tube containing 5 mL of pre-warmed DMEM containing 10% FBS.
8. Centrifuge the tube at 100 *g* for 5 min at 25 °C.
9. Discard the supernatant without disturbing the pellet.
10. Resuspend the pellet with 3 mL of fresh MEF medium.
11. Aspirate excess gelatin solution from the incubated dishes.
12. Add 4.5 mL of pre-warmed DMEM containing 10% FBS to the dish.
13. Seed the cells into gelatin-coated plates at a density of 1×10^4 cells/60-mm dish (Note 4).
14. Incubate the dishes at 37 °C with 5% CO_2, until the cells reach 80–90% confluency.
15. Keep monitoring the cells every day until use (Note 5).

10.3.3 Plating Mouse iPSCs on Feeder Cells

1. Pre-warm stem cell medium in a 37 °C water bath for 30 min (Note 6).
2. Take frozen mouse iPSCs from liquid nitrogen storage.
3. Thaw the cells quickly in a 37 °C water bath.
4. Sterilize the vial with 70% ethanol.
5. Add 5 mL of stem cell medium to a 15-mL conical tube.

6. Transfer the cells using a sterile long Pasteur pipette gently suspending into the medium in a 15-mL conical tube.
7. Shake gently the tube to suspend the cells well.
8. Centrifuge cells at 200 g for 5 min at 25 °C.
9. Aspirate the supernatant without disturbing the pellet.
10. Add 5 mL of stem cell medium.
11. Resuspend the pellet by gently pipetting up and down 2 or 3 times using a sterile long Pasteur pipette.
12. Pick up the MEF-coated dishes prepared in the previous step from the incubator.
13. Replace the MEF medium with 4 mL of iPS complete medium supplemented with 1000 U/mL of LIF.
14. Seed 0.1×10^6 of cells onto the MEF-dishes (Note 7).
15. Remove the dead cells by changing the medium on the next day.
16. Monitor the cells until the colonies reach 80% confluency.

10.3.4 Transfer Mouse iPSCs to Feeder-Less Culture

1. Cover the bottom of a dish with 2 mL sterile 0.1% gelatin (Note 8).
2. Incubate dishes for at least 30 min at 37 °C.
3. Trypsinize mouse iPSCs on MEF-dishes (Note 9).
4. Transfer the suspension into a sterile 15-mL conical tube.
5. Centrifuge cells at 100 *g* for 5 min at 25 °C.
6. Aspirate the supernatant without disturbing the pellet.
7. Resuspend gently the cell pellet in 3 mL of stem cell complete medium.
8. Aspirate excess gelatin from three dishes.
9. Add 4 mL of stem cell complete medium containing 1000 U/mL of LIF in each dish.
10. Add 1 mL of the cell suspension to each dish.
11. Incubate the dishes for 1 h at 37 °C with 5% CO_2 (Note 10).
12. Transfer the supernatant containing the mouse iPSCs to the rest of three gelatin-coated dishes.
13. Incubate the dishes at 37 °C with 5% CO_2 until stable colonies appear.
14. Change medium at 2-day intervals with stem cell medium containing 1000 U/mL of LIF.
15. Start passage when cells become 80% confident keeping the condition of iPSCs forming bright and clear colonies under inverted microscope (Note 11).

10.3.5 Evaluation of Carcinogenicity in Vitro

1. Seed miPSCs at a Density of 1×10^3 Cells/Well in 96-Well Black Plates (Corning Inc., NY) Coated with Gelatin in Stem Cell Medium Overnight (Fig. 10.2) (Note 12).

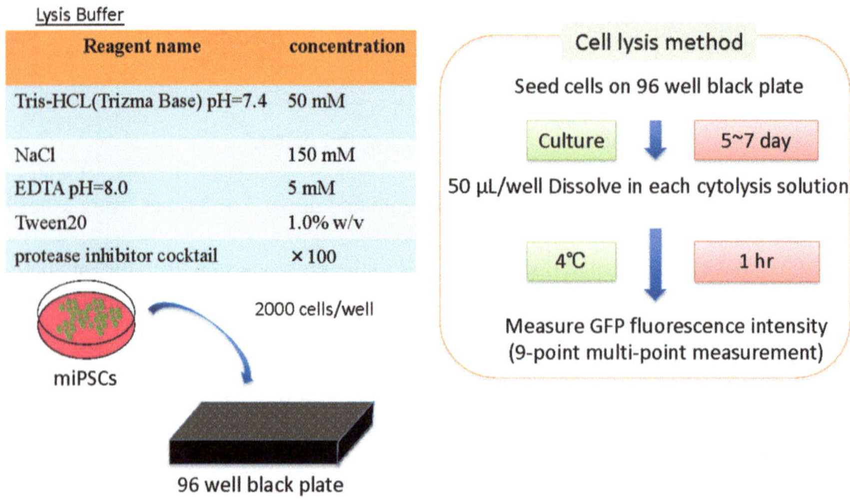

Fig. 10.2 Representative image for green fluorescent protein (GFP) Assay

2. After 24 h of incubation, replace half of the medium with fresh medium mixed with target compounds every day, up to day 6 (Note 13).
3. On day 7, wash the cells twice with PBS.
4. Add 50 μL of lysis buffer per well.
5. Incubate the plate at 37 °C for 5 min.
6. Measure the intensity of fluorescence at 509 nm which is excited at 488 nm with a fluorescence microplate reader (SH-9000 lab, Corona, Japan) (Note 14).
7. Put GFP fluorescence from the well of miPSCs cultured in CM without compounds as a control to determine the thresholds in order to distinguish the effect of the target compounds.
8. Measure the fluorescence from six wells for 5 times/well (Note: At least three independent experiments are required to calculate the averages and standard deviations for statistical analyses) (Fig. 10.3).
9. Calculate the averages and standard deviations of the readings.
10. Assess the effect of the compounds between 0 and 10 μM by the intensity of the GFP fluorescence (Fig. 10.3).
11. Determine the optimal concentration which enhances the fluorescence of GFP (Note 15).
12. Wash the wells of GFP positive cells corresponding to the results of Step 11 in 96-well clear plates with PBS and trypsinize the cells by incubating at 37 °C for 5 min.
13. Prepare a 35-mm dish coated with gelatin.
14. Pour 4 mL of DMEM containing 10% FBS into the gelatin-coated dish.
15. Transfer the cells to a gelatin-coated 60-mm dish and incubate 37 °C in 5% CO_2 up to 80% confluency with suitable medium change (Note 16).

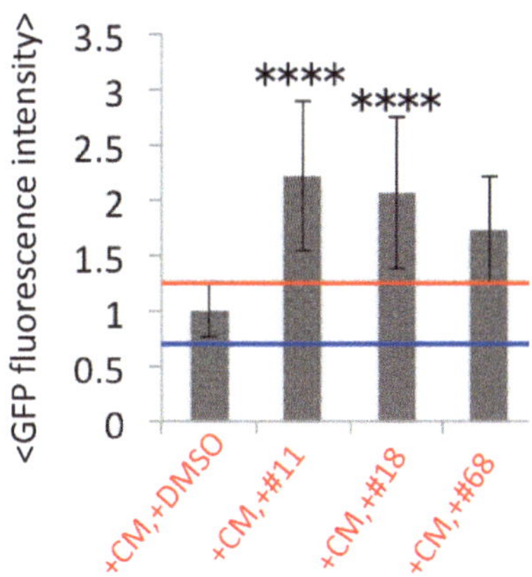

Fig. 10.3 A sample of examination to estimate the criteria. The thresholds denote average + S.D. at the top (Red) and average—S. D. at the bottom (Blue) of the control (miPSCs treated without compounds but with DMSO). In this case, DMSO is used as the solvent and should be kept less than 0.1% per well. Each bar graph is shown with the mean ± S.D. where $n \geqq 6$. $^*p < 0.01$

10.3.6 Treating CSCs with the Compounds

1. Cover the bottom of a 60-mm dish with 2 mL of sterile 0.1% gelatin (Note 17).
2. Incubate the dishes at 37 °C for at least 30 min.
3. Trypsinize miPSCs in the dishes prepared in the previous section on gelatin-coated dishes.
4. Stop the digestion with DMEM containing 10% FBS.
5. Transfer the cell suspension into a sterile 15-mL conical tube.
6. Centrifuge cells at 100 *g* for 5 min at 25 °C.
7. Aspirate the supernatant without disturbing the pellet.
8. Resuspend the cell pellet gently in 3 mL of stem cell complete medium.
9. Count the cells with a hemocytometer with trypan blue.
10. Pick up the gelatin-coated dishes from the incubator.
11. Aspirate excess gelatin from dishes.
12. Seed 0.3×10^6 cells/60-mm dish and incubate the dish at 37 C in 5% CO_2 for at least 24 h. or up to 60% confluent.
13. Change the medium to conversion medium that is containing stem cell medium and CM with or without the target compounds (Note 18).
14. Change the medium every 48 h.
15. Keep incubation at 37 °C in 5% CO_2 for 1 week.
16. Monitor the fluorescence of GFP of which expression is controlled under the Nanog promoter and take photographs of the cells at day 0, 3, and 7 with Olympus IX81 microscope equipped for fluorescence (Fig. 10.4).

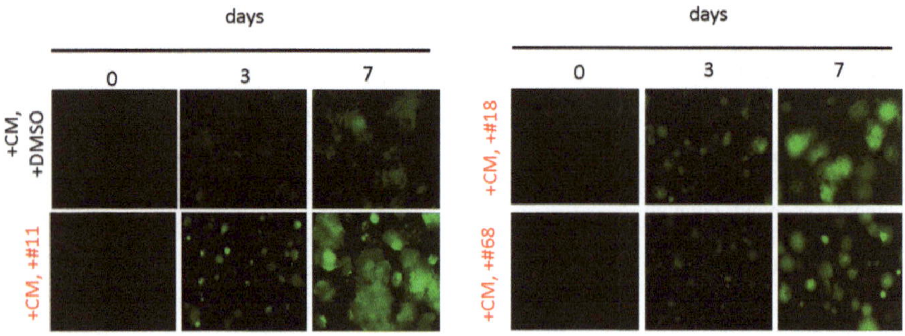

Fig. 10.4 Representative images for carcinogenic induction in the presence of chemical inhibitors #11, 18, and 68 with the CM of cancer-derived cell. DMSO is used as the solvent of the inhibitors and adjusted to be less than 0.1% in each well

10.3.7 Confirmation of Tumorigenicity by Transplantation into Nude Mice

1. Wash the GFP positive cells after 7 days from the previous section with PBS (Note 19).
2. Trypsinize the GFP-positive cells at 37 °C for 5 min.
3. Stop the digestion by adding DMEM containing 10% FBS.
4. Transfer the suspension into a sterile 15-mL conical tube.
5. Centrifuge cells at 100 g for 5min at 25 °C.
6. Aspirate the supernatant without disturbing the pellet.
7. Count the cells with hemocytometer and suspend 1×10^6 of the cells in 200 μL of HBSS.
8. Transplant 10^6 cells/mouse subcutaneously (Note 20).
9. At the end of the experiments, euthanize the mice by the isoflurane-euthanasia method (Laboratory Animal Anaesthesia (third Ed.) Paul A. Flecknell 2009). Five percent of isoflurane (DS Pharma Animal Health, Japan) was exposed to the mice and the exposure was continued until 1 min after their breathing stopped.
10. Confirm euthanasia by cervical dislocation.
11. Excise the tumor followed by paraffin embedding and sectioning.
12. Stain the sections with H&E to evaluate the malignancy (Fig. 10.5).

As a result, miPSCs treated with chemical compounds developed malignant tumors with a signature of cancer stem cells while the cells treated with only CM developed teratoma that is benign. This evaluation could successfully demonstrate that in vitro assay monitoring GFP could be a quick method to assess the carcinogenicity of compounds. We will apply this method for other compounds to assess the risk as carcinogens.

Depending on this simple method we can evaluate other chemicals such as 39, 48, 62, which show no carcinogenic effect but suppress the self-renewal (Fig. 10.6).

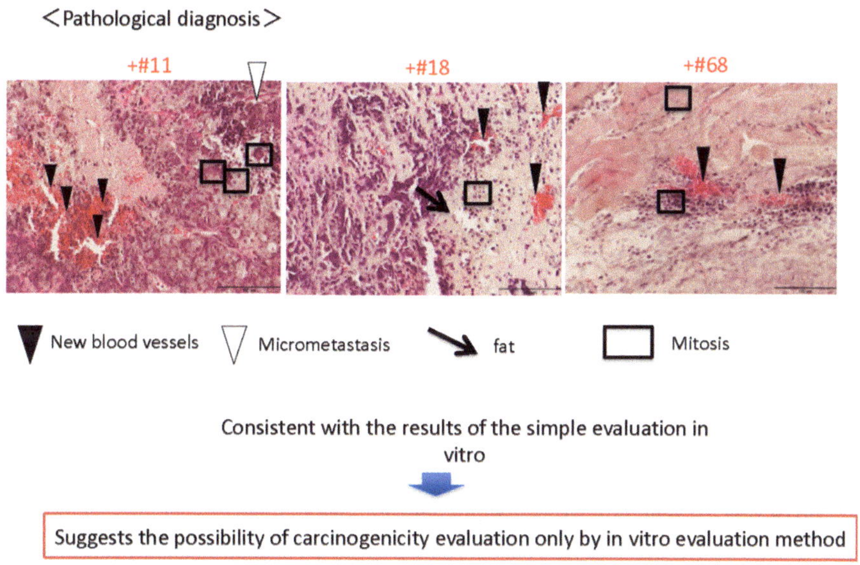

Fig. 10.5 Histopathological evaluation for the derived tumors. Scale Bar 100 μm

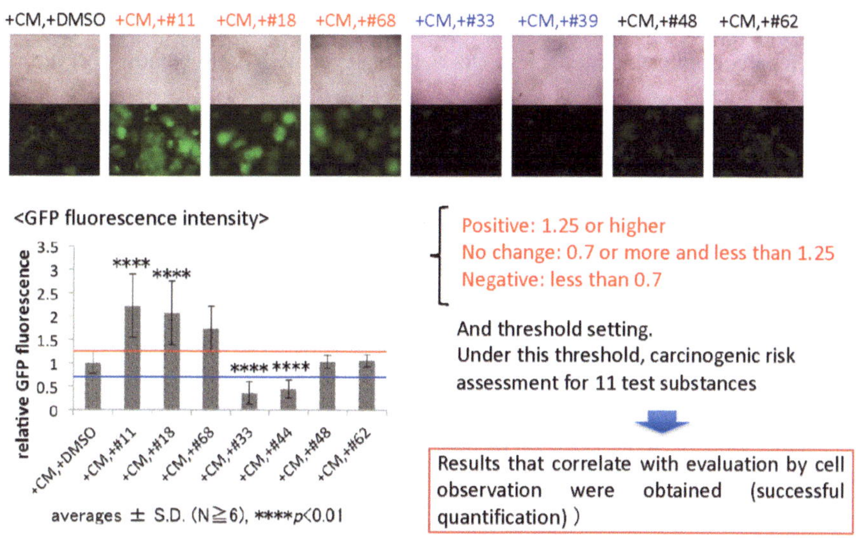

Fig. 10.6 Representative images for carcinogenic assessment in the presence of chemical inhibitors #11, 18, and 68 in comparison to #33, 39, 48, and 62, with the CM of cancer-derived cell. DMSO is used as the solvent of the inhibitors and adjusted to be less than 0.1% in each well

We demonstrated the availability of the assay monitoring GFP in miPSCs affected by the chemical compounds to assess the risk as carcinogens. The three compounds that enhanced the GFP in miPSCs in 1 week were tumorigenic in vivo.

In the following evaluation different compounds were shown to affect the GFP. Some enhanced and others attenuated the GFP. According to the results, GFP regulated with Nanog promoter in iPSCs could be proposed as an in vitro quick assay for high-throughput screening for the carcinogens.

Notes

1. Cells should be at least 80% confluent.
2. Avoid the cells from overgrowth.
3. Prepare six dishes.
4. 0.5 mL of cells from step10/dish.
5. You must use MEF-coated dishes within 2 weeks from the day of seeding and you should change medium at an interval of 4 days.
6. Do not keep the media in the water bath for more than 1 h at 37 °C as continued exposure to 37 °C may impair the components and reduce the activity of the growth factors.
7. Avoid seeding mouse iPSCs at a high density because they tend to aggregate and give rise to cells with heterogenous morphologies.
8. Prepare six dishes.
9. Do not break the colonies of iPSCs.
10. The MEF population will be attached to the bottom within 1 h.
11. Cells will be ready for conversion after two passages and will be used in the next two sections.
12. Prepare duplicates in a clear 96-well plate coated with gelatin for further experiments.
13. The concentration of the compounds should be considered beforehand.
14. The read of each well was obtained from nine points at a distance of 1 mm from the bottom of each well flashed 30 times.
15. The cells exhibiting strong GFP could be continued to grow from step 8 using the rest of the 96-well clear plates.
16. 10^6 cells/mouse should be prepared for the following section of transplantation.
17. Prepare three dishes for each condition.
18. Use the optimum concentration of the inhibitors that were assessed in Step 10 from the last section.
19. Cells from the 96-well plates and expanded to 60-mm dish will also be available.
20. Three mice were used for each compound. Measure the size every day and wait for the size to become 2000 mm^3.

References

Afify SM, Seno M. Conversion of stem cells to cancer stem cells: undercurrent of cancer initiation. Cancers (Basel). 2019;11(3):345.

Afify SM, Chen L, Yan T, Calle AS, Nair N, Murakami C, Zahra MH, Okada N, Iwasaki Y, Seno A, Seno M. Method to convert stem cells into cancer stem cells. Methods Protoc. 2019;2(3):71.

Afify SM, Sanchez Calle A, Hassan G, et al. A novel model of liver cancer stem cells developed from induced pluripotent stem cells. Br J Cancer. 2020;122:1378–90.

Afify SM, Hassan G, Seno A, Seno M. Cancer-inducing niche: the force of chronic inflammation. Br J Cancer. 2022;127(2):193–201.

Basu AK. DNA damage, mutagenesis and cancer. Int J Mol Sci. 2018;19:970.

Bonnet D, Dick JE. Human acute myeloid leukemia is organized as a hierarchy that originates from a primitive hematopoietic cell. Nat Med. 1997;13:730–7.

Calle AS, et al. A new PDAC mouse model originated from iPSCs-converted pancreatic cancer stem cells (CSCcm). Am J Cancer Res. 2016;6:2799–815.

Chen L, et al. A model of cancer stem cells derived from mouse induced pluripotent stem cells. PLoS One. 2012;7:e33544.

Collisson EA, Cho RJ, Gray JW. What are we learning from the cancer genome? Nat Rev Clin Oncol. 2012;9:621–30.

Greaves M, Maley CC. Clonal evolution in cancer. Nature. 2012;481:306–13.

Hanahan D, Weinberg RA. Hallmarks of cancer: the next generation. Cell. 2011;144:646–74.

Lapidot T, et al. A cell initiating human acute myeloid-leukemia after transplantation into SCID mice. Nature. 1994;367:645–8.

Lee HE, et al. An increase in cancer stem cell population after primary systemic therapy is a poor prognostic factor in breast cancer. Br J Cancer. 2011;104:1730–8.

Nair N, et al. A cancer stem cell model as the point of origin of cancer-associated fibroblasts in tumor microenvironment. Sci Rep. 2017;7:6838.

Wang JC. Good cells gone bad: the cellular origins of cancer. Trends Mol Med. 2010;16:145–51.

Chapter 11
Self-renewal Potential of Cancer Stem Cells

Abstract Self-renewal potential is an important characteristic for cancer stem cells (CSCs) to maintain the cancer tissues as well as for the stem cells to maintain the tissues. Depending on this potential CSCs are thought to exhibit many characteristics associated with malignancy followed by chemo- and radio-resistance due to their dormancy leading to tumor recurrence. The self-renewal in normal stem or progenitor cells appears dysregulated in CSCs resulting in the continuous expansion of tumor mass. Therefore, tumor spheres are widely used to analyze the self-renewal capability of CSCs. This method will provide a novel and practical platform for the screening of promising anti-CSC agents.

Here in the current chapter, we demonstrate the method culturing of self-renewing CSCs in tumor spheres, in suspension and in hanging drops. Furthermore, we will demonstrate the method to evaluate sphere-forming potential by the frequency as the extreme limiting dilution analysis (ELDA).

Keywords Self-renewal · Cancer stem cell · Extreme limiting dilution analysis · Hanging drop

11.1 Introduction

A stem cell divides asymmetrically or symmetrically into one or two daughter stem cells with a differentiation potential similar to the mother cell. This cell division is the process of so-called self-renewal which is unique to a stem cell (Neumüller and Knoblich 2009). The ability of self-renewal is essential for stem cells to expand their numbers during development, to maintain tissues in adults, and to restore themselves and their progenies in wound healing.

Although both processes depend on cell division, self-renewal is different from proliferation. Proliferation is a general term that describes the cell divisions in all cell types. Self-renewal is the process of dividing cells when at least one of the daughter cells has the same ability to differentiate as the mother cell. Most of the mammalian stem cells, such as hematopoietic stem cells (HSCs) and neural stem cells, maintain their multipotency by self-renewal. The regulatory mechanisms involved in the

S. M. Afify, M. Seno, *Methods in Cancer Stem Cell Biology*,
https://doi.org/10.1007/978-981-99-1331-2_11

proliferation, which many types of cells undergo, will preferentially apply to the self-renewal of stem cells (Molofsky et al. 2003; Nishino et al. 2008).

Self-renewal potential is shared between stem cell and cancer stem cells (CSCs) and is considered as one of the essential characteristics of CSCs maintaining the cancer tissues and providing the progenies of CSCs in the microenvironment (Afify and Seno 2019; Afify et al. 2022a, b).

The concept of CSC has triggered a great deal of ideas to create the novel insights to the clinical diagnosis and therapeutic strategies. A CSC model would suggest the novel way to treat cancer with agents that target the causes of CSCs and mechanisms responsible for CSCs (Afify and Seno 2019). More investigations of the self-renewal mechanisms that fuel CSC propagation will become more important to understand the functional nature of these cells and to develop novel therapeutic strategies along with the ways of identification of CSCs.

The sphere-forming assay was first introduced to confirm the self-renewal potential of stem cells (Liu 2013) and is widely used to assess the stemness of CSCs currently (Zhang et al. 2017). Anchorage-independent sphere culture is demonstrated under non-adherent and nutritionally deficient including serum-free conditions, under which differentiated tumor cells undergo apoptosis while CSCs survive, adapt, and proliferate (Liu 2013; Takai et al. 2016). This experimental approach allows the enrichment of a relatively whole subpopulation of CSCs based on the essential characteristics of CSCs regardless of their gene expression profiles. Thus, sphere formation is sometimes available as an optimal method to enrich CSCs from tumor tissues.

Our group developed CSC models from iPSCs in the cancer-inducing microenvironment (Chen et al. 2012, 2023; Nair et al. 2017; Afify et al. 2019, 2020, 2022a, b; Hassan et al. 2021; Minematsu et al. 2022).

With these CSC models, self-renewal potential has been assessed in suspension culture such as in low attachment dishes and in hanging drops. The extreme limiting dilution analysis (ELDA) will be introduced and demonstrated as the method to evaluate sphere-forming potential.

11.2 Materials

11.2.1 Reagents

- miPS-LLcm cells (Chen et al. 2012).
- miPS-MB231 cm cells (Seno Lab, Okayama, Japan).
- Corning® Matrigel®, growth factor reduced, phenol red-free (BD, cat. no. 356231).
- Dulbecco's modified Eagle's medium-high glucose (Wako, Osaka, Japan (catalogue number: 044-29765)).
- Trypsin-EDTA (0.25%) (Nacalai Tesque, Kyoto, Japan, Cat. No: 327777-44).

- Fetal bovine serum (FBS, Gibco, Life Technologies, Massachusetts, USA, Cat. No: 10437-028).
- Penicillin/streptomycin mixed solution (100 U/mL) Nacalai Tesque, Kyoto, Japan, Cat. No-26253-84).
- 70% ethanol (Sigma-Aldrich; Cat. No.: 459836-2).
- Liquid N_2 storage tank.
- Hank's balanced salt solution (HBSS) Genesee Scientific, El Cajon, USA.
- Endothelial basal medium EBM2 media (EBM-2 Single Quots Kit, Lonza, Switzerland).

11.2.2 Equipment

- Eppendorf Centrifuge 5415R, Eppendorf AG, 22331 Hamburg, Germany.
- Sanyo MCO-19AIC(UV) CO2 Incubator, Marshall Scientific, Hampton, USA.
- Type A2 Biological Safety Cabinets (E-Series).
- Olympus IX81 microscope (Olympus, Tokyo, Japan).
- Tissue culture-treated plate, 60 mm dish; TPP, AG Schaffhausen, Switzerland, Cat. No 93060;
- Tissue culture-treated plate, 100 mm dish; TPP, AG Schaffhausen, Switzerland, Cat. No 93100.
- Filter max 250 mL, TPP, Switzerland, Cat. No 99255.
- Falcon® Conical Centrifuge Tubes (15 mL; BD Falcon, New York, USA, Cat. No 352095).
- Falcon® Conical Centrifuge Tubes (50 mL; BD Falcon, New York, USA, Cat. No 352070).
- 37 °C water bath.

11.2.3 Reagent Preparation

Stem Cell Medium

Mix the following for 500 mL		
Dulbecco's modified Eagle's medium-high glucose		412 mL
FBS		75 mL
Penicillin/streptomycin mixed solution	(50 U/mL)	2.5 mL
Gibco™ L-Glutamine (200 mM)	(2 mM)	5 mL
MEM non-essential amino acids solution	(1 mM)	5 mL

11.3 Methods

11.3.1 Preparation of Gelatin-Coated Dishes

1. Transfer 0.2% gelatin solution from 4 °C to the clean bench.
2. Cover the bottom of the dish with sufficient volume of 0.2% gelatin solution.
3. Incubate the dishes for 30 min at 37 °C in an incubator with 5% CO_2.

11.3.2 Reviving CSC

1. Pick up vials of miPS-LLcm and miPS-MB231cm cells from the liquid nitrogen storage tank.
2. Thaw the cells quickly at 37 °C in a water bath.
3. Wipe the outside of the vial with 70% ethanol.
4. Add 5 mL of pre-warmed stem cell medium to a sterile 15-mL conical tube.
5. Suspend the cells gently with a sterile long Pasteur pipette.
6. Transfer the cells into the 15-mL conical tube containing the medium.
7. Centrifuge cells at 200 × g for 5 min at 25 °C.
8. Aspirate the supernatant without disturbing the pellet.
9. Add 5 mL of pre-warmed stem cell medium.
10. Resuspend the pellet by gentle pipetting up and down 2 or 3 times using a sterile long Pasteur pipette.
11. Pick up the gelatin-coated dishes prepared in the previous step from the incubator.
12. Aspirate the excess gelatin.
13. Add 4 mL of iPS medium.
14. Seed the cells on two gelatin-coated dishes dividing the cells into high and low number.
15. Remove the dead cells by changing medium on the next day.
16. Monitor the cells and take photos every day.
17. After three passages from reviving, the cells are ready for further experiments.

11.3.3 Sphere Formation Protocol

1. Detach the cells of CSCs with 0.05% trypsin by incubation for 5 min at 37 °C (Note 1).
2. Stop trypsinization by adding pre-warmed medium containing at least 5% FBS.
3. Centrifuge the cell suspension for 5 min at 25 °C at 200 × g.
4. Aspirate the supernatant without disturbing the pellet.
5. Resuspend the cells in 1 mL of pre-warmed medium.
6. Count the cells using a hemocytometer.

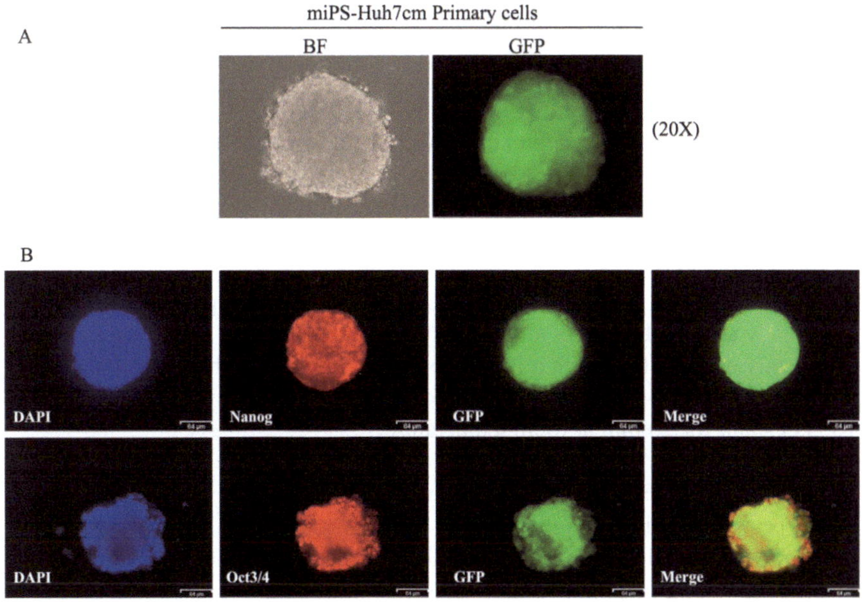

Fig. 11.1 (**a**) Sphere-formation assay in a serum-free medium shows spherogenic potential of miPS-Huh7cm primary cells. (**b**) Immunofluorescence staining for Nanog and Oct 3/4 in tumor spheroid derived from miPS-Huh7cm primary cells. Scale bars represent 64 μm

7. Adjust cell concentration to 1×10^6 cells/mL in spheroid medium (Note 2).
8. Seed 4×10^4 cells in a 60-mm dish with low attachment hydrophobic surface for appropriate suspension of spheres.
9. Incubate the dish at 37 °C in a 5% CO_2 humidified incubator for 7 days (Note 3).
10. Change a half volume of the culture medium with fresh spheroid medium every 3 days.
11. Take pictures of the sphere using Olympus IX81 microscope equipped with fluorescence (Fig. 11.1a).

11.3.4 Immunofluorescence Staining

1. Coat a coverslip with Matrigel in a 12-well plate (Note 4).
2. Incubate the plate for 30 min at 37 °C in a humidified incubator with an atmosphere of 5% CO_2.
3. Collect the spheres by transferring the cells and media into 15-mL conical tubes using a serological pipette from Step 10 of the previous section.
4. Allow the spheres to sediment by gravity for 10 min at room temperature.

5. Aspirate the supernatant leaving approximately 200 μL with the tumor spheres in the conical tube (Note 5).
6. Add 1 mL of PBS and softly suspend.
7. Allow the spheres again to sediment by gravity for 10 min.
8. Aspirate PBS leaving approximately 200 μL with the tumor spheres in the conical tube.
9. Add 200 μL of sphere medium to the spheres.
10. Transfer the tumor spheres onto the coverslip coated with Matrigel coated in a 12-well plate.
11. Let stand for 1 h until the spheres attached to the coverslip.
12. Gently add 1 mL of PBS to the well to rinse the spheres.
13. Gently aspirate PBS with a Pasteur pipette without disturbing the spheres.
14. Add 400 μL of 4% paraformaldehyde to fix the spheres.
15. Let stand for 20 min at room temperature.
16. Aspirate paraformaldehyde solution.
17. Add 1 mL of PBS, stand for 5 min and aspirate PBS.
18. Repeat the previous step five times.
19. Add 400 μL of 0.5% Triton X-100 in PBS and stand for 30 min at room temperature to permeabilize the fixed spheres.
20. Add 1 mL of PBS to the fixed and permeabilized spheres, stand for 5 min and aspirate PBS.
21. Repeat the previous step 3 times.
22. Incubate the spheres in blocking buffer (0.1% bovine serum albumin (BSA) in PBS) for at least 1 h at room temperature.
23. Add 1 mL of PBS, stand for 5 min and aspirate PBS.
24. Repeat the previous step five times.
25. Add 400 μL of the primary antibody (anti-Nanog or anti-Oct3/4 antibody) appropriately diluted in blocking buffer to the spheres and incubate overnight at 4 °C.
26. Aspirate the solution of the primary antibody after incubation.
27. Add 1 mL of PBS, stand for 5 min and aspirate PBS.
28. Repeat the previous step five times.
29. Add 400 μL of PBS containing the secondary antibody linked with Alexa-555 to the spheres and incubate for 1 h at room temperature (Note 6).
30. Aspirate the solution of the secondary antibody after incubation.
31. Add 1 mL of PBS, stand for 5 min and aspirate PBS.
32. Repeat the previous step three times.
33. Pick up the cover slip using forceps.
34. Put the coverslip on a slide glass with 20 μL of mounting medium containing 4',6-diamidino-2-phenylindole (DAPI).
35. Gently invert the coverslip on the slide not to trap air bubbles.
36. Stand for 2 h at 25 °C to dry the mounting medium.
37. Observe the slide and take photographs of stained spheres under a fluorescent microscope (Fig. 11.1b).

11.3.5 Passage Spheres

1. Collect the spheres after 7 days by transferring the cells and media into 15-mL conical tubes using a serological pipette.
2. Allow the spheres to sediment by gravity for 10 min at 25 °C.
3. Aspirate the supernatant leaving approximately 200 μL in the conical tube.
4. Add 500 μL 0.05% trypsin in PBS to the conical tube.
5. Incubate the conical tube at 37 °C for 5 min.
6. Stop trypsinization by adding 1 mL of pre-warmed media containing 10% FBS.
7. Centrifuge the cells at 200 × g for 5 min at 25 °C.
8. Count the cells using a hemocytometer.
9. Adjust the cell concentration to 1×10^6 cells/mL in a spheroid medium.
10. Seed 4×10^4 cells in a 60-mm dish with low attachment hydrophobic surface for appropriate suspension of spheres.
11. Incubate the dish at 37 °C in a 5% CO_2 humidified incubator for 7 days.
12. Change a half volume of the culture medium with fresh spheroid medium every 3 days.

11.3.6 Hanging Drop Sphere Formation

To further assess the self-renewal potential of CSCs, a hanging drop method could be used to produce spheroids. Fixed numbers of cells in 20-μL drops (1–1000 cells/drop) are placed on the inside of the lid of the 60-mm dish. Then the bottom of the dish is filled with PBS so as not to get the spheroid drops dried during the incubation (Fig. 11.2). In this method, the cells cultured in the hanging drop are designed to grow without anchorage to begin aggregation. The spheres will become visible within the first 24 h when the initial number of the cells is more than 500 per drop.

1. Adjust the cell from Step 6 of the first section "Sphere Formation Protocol" to the concentration of 1×10^4 cells/mL in spheroid medium (Note 7).
2. Open the lid from a 60-mm tissue culture dish and place 5 mL of PBS at the bottom of the dish (Note 8).
3. Invert the lid and deposit 20-μL drops to the inside of the lid using a 20-μL pipette (Note 9).
4. Return the lid to the dish filled with PBS.
5. Incubate the dish at 37 °C in a 5% CO_2 incubator for at least 7 days.
6. Monitor the spheres in the drops daily and incubate until aggregates are visible (Note: After 7 days, you will be able to see the spheroid by eyes).
7. Invert the lid.
8. Add from 2 mL of pre-warmed medium and gently mix the drops of spheres. (Note: Just swirl gently without breaking the sphere but do not use pipettes).
9. Transfer the supernatant containing spheres to the bottom of new 60-mm dish using serological pipettes.

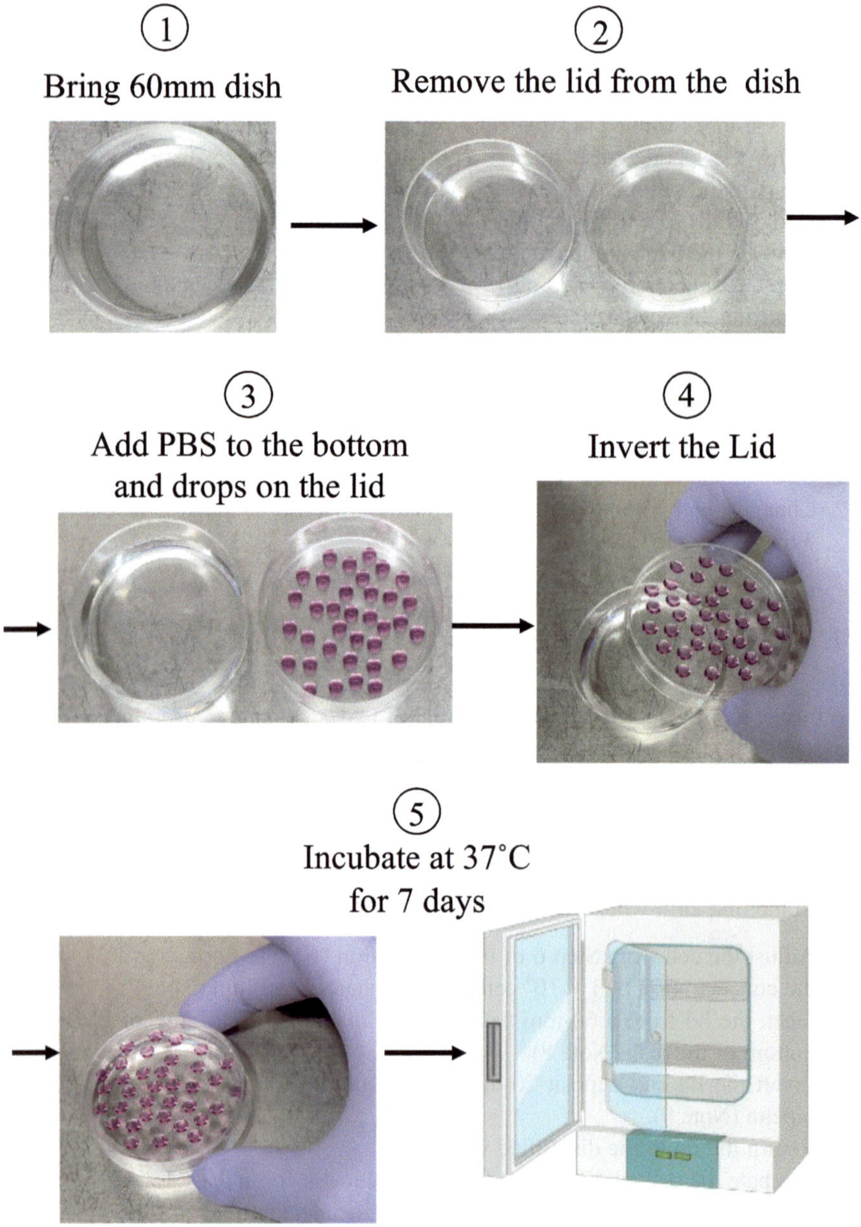

Fig. 11.2 A schematic diagram showing the steps for forming spheres using the hanging drop method. The inverted lids of culture dishes are incubated for periods of 24–96 h after 20 drops of cell suspension are placed on them

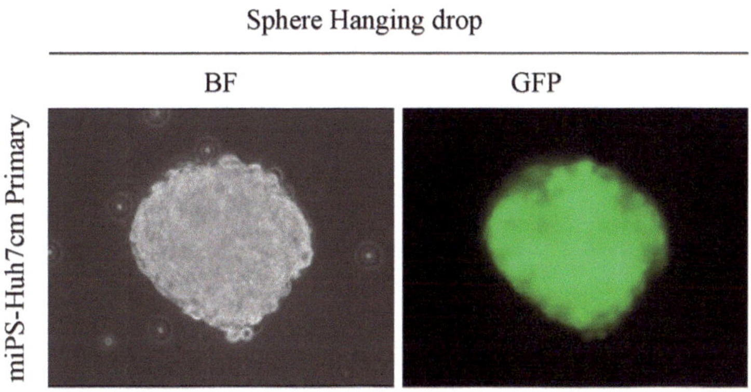

Fig. 11.3 Images of CSCs incubated in hanging drop culture for 7 days

10. Observe the spheres and take photographs of spheres under a fluorescent microscope (Fig. 11.3).

11.3.7 Extreme Limiting Dilution Assay (ELDA)

The in vitro limiting dilution assay is performed by counting the frequency of spheres of CSC in suspension culture.

1. Grow cells to confluence in a 60-mm dish at 37 °C in an incubator with 5% CO_2.
2. Dissociate CSCs with trypsin into a single-cell suspension by incubating at 37 °C for at least 3 min until the cell becomes round (Note 10).
3. Stop trypsinization by adding a pre-warmed medium containing 10% FBS.
4. Centrifuge cells at 200 × g for 5 min at 25 °C.
5. Remove the supernatant without disturbing the pellet.
6. Resuspend the cells in 1 mL pre-warmed sphere medium (Note 11).
7. Determine the cell number using hemocytometer.
8. Prepare serially diluted cells in tubes containing the following final cell concentration per tube (Note: You can start from 50 to 200 cells/well depending on the character of cells). The following tubes are designed for one 96-well ultra-low attachment plate starting from 50 cells/well with five twofold dilutions. Prepare enough volume of solution for each tube. If you are providing 100 μL/well, you should prepare 1.8 mL but not just 1.6 mL.

   A. Tube 1 (for 50 cells/100 μL/well × 16 wells).
   B. Tube 2 (for 25 cells/100 μL/well × 16 wells).
   C. Tube 3 (for 12 cells/100 μL/well × 16 wells).
   D. Tube 4 (for 6 cells/100 μL/well × 16 wells).
   E. Tube 5 (for 3 cells/100 μL/well × 16 wells).
   F. Tube 6 (for 1.5 cells/100 μL/well × 16 wells).

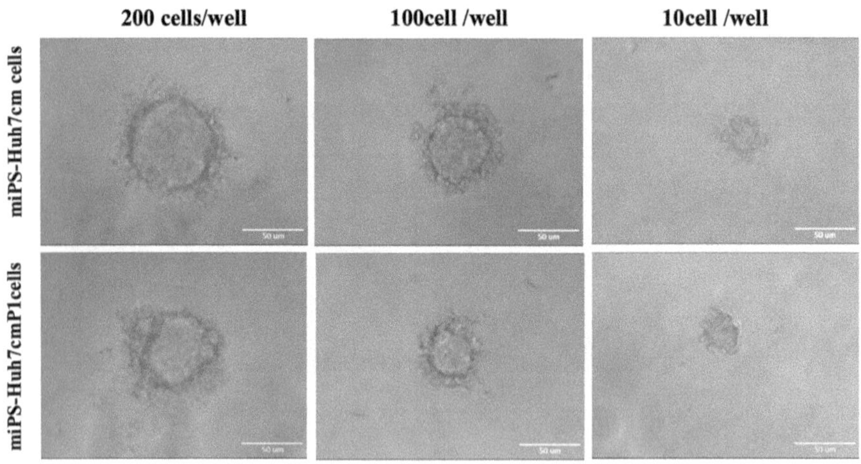

Fig. 11.4 Representative images for spheroid culture of miPS-Huh7cm and miPS-Huh7cmP1 cells with limiting dilution assay

9. Seed the cells into a 96-well ultra-low attachment plate, in the presence of sphere media (Note 12).
10. Make 8 replicates of wells or more up to 96 for one "cells/well" (Note 13).
11. Incubate in a 37 °C, 5% CO_2 humidified incubator for 7–10 days.
12. Monitor each well for signs of spheroid formation every day.
13. Count the number of all wells that contain one or more spheres greater than ~50 μm diameter at day 10 (Note 14).
14. Take pictures for developed spheres at different cell numbers (Fig. 11.4).
15. Input the number of wells with spheres into ELDA software at http://bioinf.wehi.edu.au/software/elda/.
16. Select "Plot Results" checkbox; default confidence interval is set at 0.95 and click "Run" (Note: The results plot can be printed as a PDF file).
17. Check the results as shown in Tables 11.1, 11.2, 11.3, and Fig. 11.5.

Notes

1. The cells should be 80–90% confluent and in good condition from the previous section.
2. Cells can be kept on ice up to 1 h until use.
3. Avoid the incubation for long until the center of spheroids starts to get dark.
4. This coverslip will be used at Step 10.
5. Be careful not to aspirate the tumor spheres.
6. All steps following this step should be performed in the dark.
7. Cells can be kept on ice up to 1 h until use.
8. This is very important because this water will act as a hydration chamber.

Table 11.1 List for limiting dilution data entered to software http://bioinf.wehi.edu.au/software/elda/

Counter	Dose	Tested	Response	Group
1	1	96	0	A
2	10	96	25	A
3	100	96	80	A
4	200	96	96	A
5	1	96	0	B
6	10	96	38	B
7	100	96	96	B
8	200	96	96	B

Table 11.2 Confidence intervals for 1 (stem cell frequency)

Group	Lower	Estimate	Upper
A	56.9	46.9	38.7
B	26.4	20.4	15.8

Table 11.3 Overall test for differences in stem cell frequencies between any of the groups

Chisq	DF	P.value
25.8	1	3.73e-07

9. Make sure that drops are placed apart so as not to disturb one another. Place approximately 30 drops per dish. The maximum volume of each drop should be less than 25 μL.

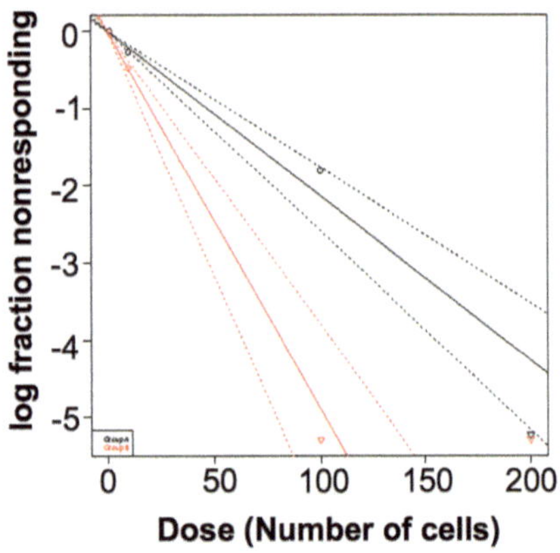

Fig. 11.5 Extreme limiting dilution assay assessment of the limiting dilution sphere-forming potential of miPS-Huh7cm (black line) and miPS-huh7cmP1 cells (red line)

10. There is no need to wait until the cells are detached from the bottom of the dish so as to avoid cell death increasing by the incubation with trypsin.
11. Make sure cells are properly dissociated without aggregation.
12. You may want to confirm the number of the cells/well in C and D by a microscope after seeding.
13. You can choose an appropriate number of replicates depending on the character of the cells.
14. Sphere size is measured using pre-calibrated scale bar function in the image capture software.

References

Afify SM, Seno M. Conversion of stem cells to cancer stem cells: undercurrent of cancer initiation. Cancers (Basel). 2019;11(3):345.

Afify SM, Hassan G, Osman A, Calle AS, Nawara HM, Zahra MH, El-Ghlban S, Mansour H, Alam MJ, Abu Quora HA, Du J, Seno A, Iwasaki Y, Seno M. Metastasis of cancer stem cells developed in the microenvironment of hepatocellular carcinoma. Bioengineering (Basel). 2019;6(3):73.

Afify SM, Sanchez Calle A, Hassan G, Kumon K, Nawara HM, Zahra MH, Mansour HM, Khayrani AC, Alam MJ, Du J, Seno A, Iwasaki Y, Seno M. A novel model of liver cancer stem cells developed from induced pluripotent stem cells. Br J Cancer. 2020;122(9):1378–90.

Afify SM, Hassan G, Seno A, Seno M. Cancer-inducing niche: the force of chronic inflammation. Br J Cancer. 2022a;127(2):193–201.

Afify SM, Hassan G, Yan T, Seno A, Seno M. Cancer stem cell initiation by tumor-derived extracellular vesicles. Methods Mol Biol. 2022b;2549:399–407.

Chen L, Kasai T, Li Y, et al. A model of cancer stem cells derived from mouse induced pluripotent stem cells. PLoS One. 2012;7(4):e33544.

Chen L, Liu Y, Xu Y, Afify SM, Gao A, Du J, Liu B, Fu X, Liu Y, Yan T, Zhu Z, Seno M. Up-regulation of Dsg2 conferred stem cells with malignancy through wnt/β-catenin signaling pathway. Exp Cell Res. 2023;422(1):113416.

Hassan G, Afify SM, Du J, Nawara HM, Sheta M, Monzur S, Zahra MH, Abu Quora HA, Mansour H, El-Ghlban S, Uesaki R, Seno A, Seno M. MEK1/2 is a bottleneck that induces cancer stem cells to activate the PI3K/AKT pathway. Biochem Biophys Res Commun. 2021;583:49–55.

Liu S, Li N, Yu X, et al. Expression of intercellular adhesion molecule 1 by hepatocellular carcinoma stem cells and circulating tumor cells. Gastroenterology. 2013;144:1031–41.

Minematsu H, Afify SM, Sugihara Y, Hassan G, Zahra MH, Seno A, Adachi M, Seno M. Cancer stem cells induced by chronic stimulation with prostaglandin E2 exhibited constitutively activated PI3K axis. Sci Rep. 2022;12(1):15628.

Molofsky AV, Pardal R, Iwashita T, Park IK, Clarke MF, Morrison SJ. Bmi-1 dependence distinguishes neural stem cell self-renewal from progenitor proliferation. Nature. 2003;425 (6961):962–7.

Nair N, Calle AS, Zahra MH, Prieto-Vila M, Oo AKK, Hurley L, Vaidyanath A, Seno A, Masuda J, Iwasaki Y, Tanaka H, Kasai T, Seno M. A cancer stem cell model as the point of origin of cancer-associated fibroblasts in tumor microenvironment. Sci Rep. 2017;7(1):6838.

Neumüller RA, Knoblich JA. Dividing cellular asymmetry: asymmetric cell division and its implications for stem cells and cancer. Genes Dev. 2009;23(23):2675–99.

Nishino J, Kim I, Chada K, Morrison SJ. Hmga2 promotes neural stem cell self-renewal in young but not old mice by reducing p16Ink4a and p19Arf expression. Cell. 2008;135(2):227–39.

Takai A, Fako V, Dang H, et al. Three-dimensional Organotypic culture models of human hepatocellular carcinoma. Sci Rep. 2016;6:21174.

Zhang Y, Xu W, Guo H, et al. NOTCH1 signaling regulates self-renewal and platinum Chemoresistance of cancer stem-like cells in human non-small cell lung cancer. Cancer Res. 2017;77:3082–91.

Chapter 12
Differentiation Potential of Cancer Stem Cells In Vitro

Abstract Cancer stem cells play an indispensable role in the development and progression of cancer forming the tumor microenvironment. This microenvironment is composed of a diversity of cell types such as endothelial cells, fibroblasts, immune cells, and so on. These cells cooperate together in the extracellular matrix initiating tumors and enhancing progression. Cancer stem cells have differentiation abilities to generate the tissue-specific original lineage that is similar to their normal stem cell counterparts. In this context, cancer stem cells also have the potential to transdifferentiate into vascular endothelial cells and pericytes, cancer-associated fibroblasts, and hematopoietic stem cells indicating that the progenies of cancer stem cells can form the tumor microenvironment. Although these components in the tumor stroma may explain the heterogeneity of the tumor, many issues related to this potential are yet uncovered. In this chapter, we summarize the multi-lineage differentiation and trans differentiation potentials of cancer stem cells into different tumor stromal cells.

Keywords Differentiation · Endothelial cells · Adipocytes · Cancer stem cells

12.1 Introduction

In CSCs, cytoplasmic signaling pathways such as Notch, Hedgehog, and Wnt are constitutively active regulating differentiation and self-renewal in a similar way of normal stem cells (Takebe et al. 2015; Chen et al. 2022). Simultaneously, the niches of CSCs provide the cues to essentially regulate and maintain the cellular hierarchy just like normal stem cells. CSCs will not be able to survive maintaining self-renewal capacity and specific lineage without their niches (Yan et al. 2014, Afify and Seno 2019, Afify et al. 2022a). CSCs are potent to differentiate into different lineage progeny cells (Hassan et al. 2019, Osman et al. 2020a, b, Kumon et al. 2021, Nawara et al. 2021) and are apparently responsible for tumor growth and metastasis (Afify et al. 2019, Mansour et al. 2020, Mansour et al. 2022, Afify et al. 2022b). The differentiation of CSCs have been described in numerous solid tumors including pancreatic (Li et al. 2007), prostate (Collins et al. 2005), lung (Eramo et al. 2008), and liver cancers (Ma et al. 2007, Lee et al. 2011). Al-Hajj et al. demonstrated that

S. M. Afify, M. Seno, *Methods in Cancer Stem Cell Biology*,
https://doi.org/10.1007/978-981-99-1331-2_12

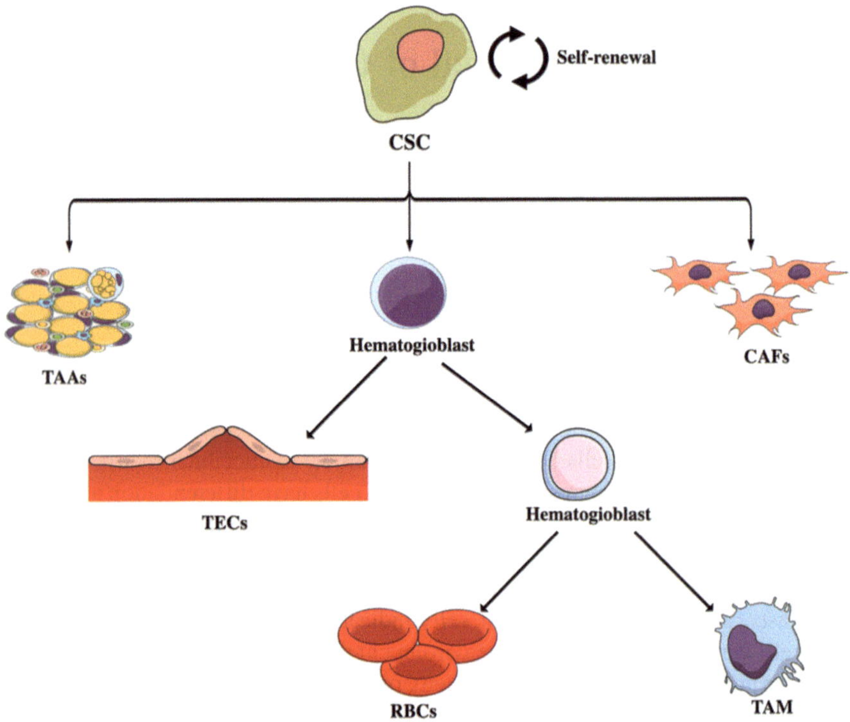

Fig. 12.1 Schematic representation of cancer stem cells' differentiation into tumor stromal cells, tumor-associated macrophages (TAM), tumor-associated fibroblasts (CAF), tumor-associated endothelial cells (TEC), erythrocytes, and tumor-associated adipocytes (TAA)

CSCs could differentiate into non-tumorigenic phenotypes (Al-Hajj et al. 2003). Collectively, CSCs exhibit differentiation in vitro and tumorigenesis developing the organization of cellular hierarchy in vivo (Fig. 12.1).

A cancer-associated adipocyte (CAA) has been identified as an adipose-like tissue in the tumor microenvironment (TME) (Dirat et al. 2011). CAA exhibits some characteristics, such as dispersed lipid droplets, fibroblast-like phenotypes, relatively small in size, high expression of collagen VI, and low expression of adiponectin (APN) and other adipokines. CAAs secrete, furthermore, chemokine (C–C motif) ligand 2 (CCL2), chemokine (C–C motif) ligand 5 (CCL5), interleukin-1β (IL-1β), interleukin-6 (IL-6), tumor necrosis factor-alpha (TNF-α), vascular endothelial growth factor (VEGF), and leptin (Fujisaki et al. 2015, D'Esposito et al. 2016). Although the origin of CAAs is still unclear, CAAs in a tumor tissue as well as in TME could be attributed to the proliferation and invasion of tumor cells into adjacent tissues (Cozzo et al. 2017). Yan et al. reported that CSCs could give rise to CAAs (Yan et al. 2014).

Endothelial cells (ECs) play important roles in vascular homeostasis interacting with blood cells as well as the cells present in the vessel structures. They are responsible for the adhesion of platelets and thrombosis, immune responses, inflammation, blood flow, etc. (Esper et al. 2006; Michiels 2003). While vascular vessels are the most available sources of ECs for experiments in vitro and in vivo, solid tumors exhibit diverse patterns of neovascularization such as sprouting and intussusception angiogenesis, vessel co-option, vasculogenesis, vascular mimicry, CSC transdifferentiation, and so on (Jain and Carmeliet 2012). Vascular mimicry is a process in which tumor cells incorporate into blood vessels to form a vascular structure that is similar to normal vessels without ECs where the staining is positive for periodic acid-Schiff (Weis and Cheresh 2011, Wagenblast et al. 2015). CSCs may directly contribute to the appearance of endothelial cells in tumor vessels and the vascular mimicry could be a mid-stage of EC maturation implying the endothelial differentiation of CSCs. This differentiation was further supported by the studies demonstrating that CSCs in multiple tumors enriched with vascular structures as the result of tumor angiogenesis (Prieto-Vila et al. 2016).

12.2 Materials

12.2.1 Reagents

- miPS-LLcm cells (Seno Lab, Okayama, Japan).
- miPS-MB-231 cm cells (Seno Lab, Okayama, Japan).
- Corning® Matrigel®, growth factor reduced, phenol red-free (BD, cat. no. 356231).
- Dulbecco's modified Eagle's medium-high glucose (Wako, Osaka, Japan (catalog number: 044–29,765)).
- Trypsin-EDTA (0.25%) (Nacalai Tesque, Kyoto, Japan, Cat. No: 327777–44).
- Fetal bovine serum (FBS, Gibco, Life Technologies, Massachusetts, USA, Cat. No: 10437–028).
- Penicillin/streptomycin mixed solution (100 U/mL) Nacalai Tesque, Kyoto, Japan, Cat. No-26253-84).
- 70% ethanol (Sigma-Aldrich; Cat. No.: 459836-2).
- Liquid N_2 storage tank.
- Hank's balanced salt solution (HBSS) Genesee Scientific, El Cajon, USA.
- Endothelial basal medium EBM2 media (EBM-2 SingleQuots Kit, Lonza, Switzerland).

12.2.2 Equipment

- Eppendorf Centrifuge 5415R, Eppendorf AG, 22331 Hamburg, Germany.
- Sanyo MCO-19AIC(UV) CO2 Incubator, Marshall Scientific, Hampton, USA.
- Type A2 Biological Safety Cabinets (E-Series).

- Olympus IX81 microscope (Olympus, Tokyo, Japan).
- Tissue culture-treated plate, 60 mm dish; TPP, AG Schaffhausen, Switzerland, Cat. No 93060.
- Tissue culture-treated plate,100 mm dish; TPP, AG Schaffhausen, Switzerland, Cat. No 93100.
- Filter max 250 mL, TPP, Switzerland, Cat. No 99255.
- Falcon® Conical Centrifuge Tubes (15 mL; BD Falcon, New York, USA, Cat. No 352095).
- Falcon® Conical Centrifuge Tubes (50 mL; BD Falcon, New York, USA, Cat. No 352070).
- 37 °C water bath.

12.2.3 Reagent Preparation

12.2.3.1 Stem Cell Medium

Mix the following for 500 mL:		
Dulbecco's modified Eagle's medium-high glucose		412.5 mL
FBS		75 mL
Penicillin/streptomycin mixed solution	(50 U/mL	2.5 mL
Gibco™ L-glutamine (200 mM)	(2 mM)	5 mL
MEM non-essential amino acids solution	(1 mM)	5 mL

12.2.3.2 Tube Medium

Endothelial basal medium EBM2 med	
Human epidermal growth factor	(5 ng/mL)
Vascular endothelial growth factor	(VEGF; 0.5 ng/mL)
R3-insulin-like growth factor-1	(20 ng/mL)
Ascorbic acid	(1 μg/mL).
Hydrocortisone	(0.2 μg/mL)
Human basic fibroblast growth factor	(FGF; 10 ng/mL).
Heparin	(22.5 μg/mL)
FBS	(0.02 mL/mL)

12.3 Methods

12.3.1 Preparation of Gelatin-Coated Dishes

1. Transfer 0.2% gelatin solution from 4 °C to the clean bench.
2. Cover the bottom of the dish with sufficient volume of 0.2% gelatin solution. Note 1.
3. Incubate the dishes for 30 min at 37 °C in an incubator with 5% CO_2.

12.3.2 Reviving CSC

1. Pick up vials of miPS-LLcm and miPS-MB231cm cells from the liquid nitrogen storage tank.
2. Thaw the cells quickly at 37 °C in a water bath.
3. Wipe the outside of the vial with 70% ethanol.
4. Add 5 mL of pre-warmed stem cell medium to a sterile 15-mL conical tube.
5. Suspend the cells gently with a sterile long Pasteur pipette.
6. Transfer the cells into the 15-mL conical tube containing the medium.
7. Centrifuge cells at 200 × g for 5 min at 25 °C.
8. Aspirate the supernatant without disturbing the pellet.
9. Add 5 mL of pre-warmed stem cell medium.
10. Resuspend the pellet by gently pipetting up and down 2 or 3 times using a sterile long Pasteur pipette.
11. Pick up the gelatin-coated dishes prepared in the previous step from the incubator.
12. Aspirate the excess gelatin.
13. Add 4 mL of iPS medium.
14. Seed the cells on two gelatin-coated dishes dividing the cells into high and low number.
15. Remove the dead cells by changing medium on the next day.
16. Monitor the cells and take photos every day (Fig. 12.2a).
17. After three passages from reviving, the cells are ready for further experiments.

12.3.3 Endothelial Differentiation

As CSCs are thought to differentiate into platelet-endothelial cell adhesion molecule-1 (CD31) positive endothelial cells, miPS-LLcm cells in adhesive culture were also examined for CD31 expression by immunofluorescence (Fig. 12.2b). Based on the significant CD31 expression discovered, miPS-LLcm cells were supposed to be capable of differentiating into endothelial cells. Thus, in the

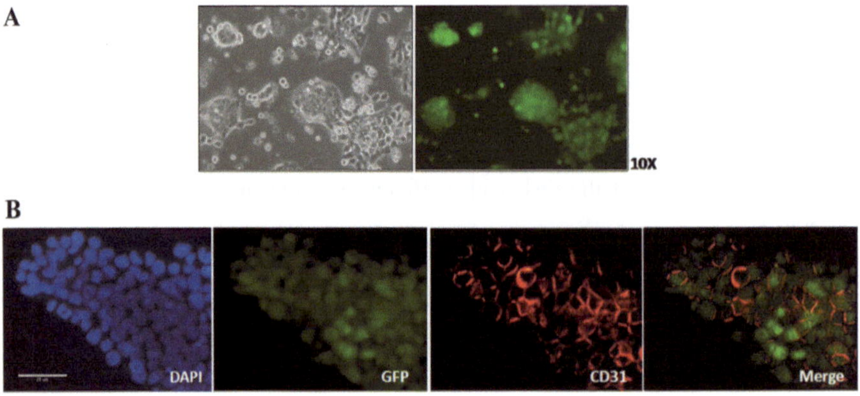

Fig. 12.2 (**a**) Representative image for miPS-LLcm cells. (**b**) Representative images of miPS-LLcm cells stained with anti-CD31 antibody (middle right) and DAPI (left). Green fluorescence (middle left). Merge of green fluorescence and immunoreactive CD31 (right)

following steps, we will further assess the ability to form tube-like structures when assessed on Matrigel with GFP, as well as the expression of CD31 in the formed tube by immunofluorescence.

12.3.3.1 Tube Formation Assay

1. Pick up the bottle of Matrigel at −20 °C in a freezer.
2. Place the bottle in a refrigerator at 4 °C overnight.
3. Place a coverslip in a well of 12-well plate.
4. Pour Matrigel after melting on the cover slip in a well.
5. Incubate the plate at 37 °C in an incubator with 5% CO_2 for more than 30 min.
6. Take the 60-mm dish of CSCs at 80% confluent out from the CO_2 incubator.
7. Aspirate the medium from the 60-mm dish.
8. Add 2 ml 0.025% trypsin-EDTA solution per dish.
9. Incubate the dish at 37 °C until the cell shape becomes round. Note 2.
10. Pour 5 ml of warm medium in the dish.
11. Transfer the cell suspension into a sterile 15-ml conical tube.
12. Centrifuge the tube at 200 × g for 5 min at 25 °C.
13. Discard the supernatant and suspend the cell pellet in a basal medium EBM-2.
14. Count cell number with a hemocytometer.
15. Prepare 5.0×10^5 cells suspended in 1 mL of EBM-2 medium in a 1.5-mL tube.
16. Seed the cells on the cover slip coated with Matrigel in a well.
17. Incubate 12-well plate at 37 °C in the incubator with 5% CO_2 for 6–16 h.
18. Photograph the tubular network formed in the wells using a digital camera attached to an inverted microscope with 10X or 20X objective lens (Fig. 12.3a).

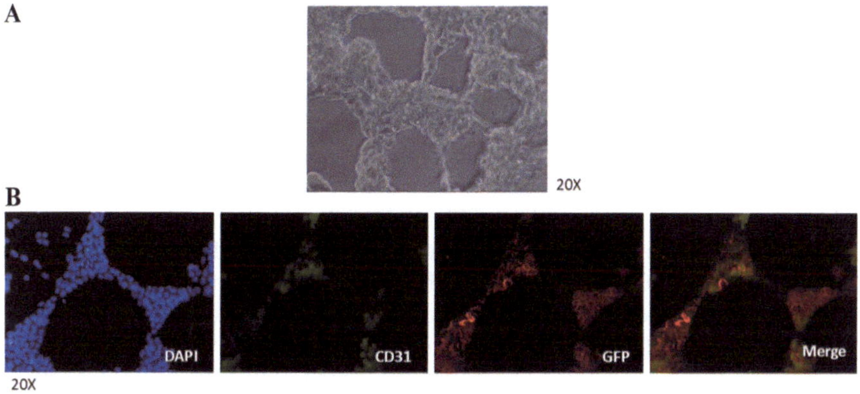

Fig. 12.3 (**a**) Representative image for miPS-LLcm cell-derived tubes. Lens (20X). (**b**) Representative images of miPS-LLcm cells forming tube structure on Matrigel stained with anti-CD31 antibody (middle right) and DAPI (left). Green fluorescence (middle left). Merge of green fluorescence and immunoreactive CD31 (right)

12.3.3.2 Immunostaining against CD31

1. Aspirate the medium from the wells.
2. Rinse the wells with PBS.
3. Aspirate PBS.
4. Fix the samples with 4% paraformaldehyde in PBS for 15 min at 25 °C. Note: Alternatively, add 500 μL of −20 °C cold methanol and incubate the plate for 10 min at 25 °C.
5. Aspirate fixative after incubation.
6. Rinse the tubes with PBS for 5 min and repeat the wash three times.
7. Block the plate with 10% BSA in PBS for 30 min at 25 °C.
8. Aspirate the buffer.
9. Add anti-CD31 antibody with appropriate dilution in PBS.
10. Incubate the plate at 4 °C overnight or at 37 °C for 1 h.
11. Rinse the plate with PBS for 5 min and repeat the wash three times.
12. Incubate the samples with the secondary antibody conjugated with fluorescent dye at appropriate dilution in PBS at 25 °C for 1 h in a dark place.
13. Rinse the plate with PBS for 5 min and repeat the wash three times. Note 3.
14. Mount the samples on slide glass with a drop of mounting medium containing DAPI.
15. Photograph the tubular network formed on the cover slip using a fluorescent microscope with ×4 or ×10 objective equipped with a digital camera (Fig. 12.3b).

12.3.4 Adipogenic Differentiation and Nile Red Staining

1. Passage the CSCs with a low number of approximately 1.0×10^5 cells/60-mm dish when the density becomes 80% confluent after the third culture.
2. Remove the medium carefully from the 60-mm dish when the cell density reaches 80% confluent again.
3. Remove dead cells by washing with fresh medium.
4. Add 4 mL/60-mm dish of the adipogenic medium (DMEM containing 10% FBS, 0.5 mM of 3-Isobutyl-1-methylxanthine (Wako, Japan), 1.0 μM of dexamethasone (Wako, Japan), and 1.0 μg/mL of insulin (Wako, Japan).
5. Keep culturing for 2 days at 37 °C in an incubator with 5% CO_2.
6. Change medium at day 3 to DMEM containing 10% FBS and 1.0 μg/mL of insulin.
7. After the first day, change the media every two days until the seventh day.
8. Stain the cells at day 7 with Nile Red (Wako, Japan) for the analysis of lipid localization.
9. Remove the medium carefully at day 7.
10. Wash the dish with PBS.
11. Add fresh medium supplemented with 0.1% Nile Red.
12. Incubate the dish at 37 °C in an incubator with 5% CO_2 for 5 min.
13. Take the dish out from the incubator after 5 min.
14. Wash the dish with PBS carefully.
15. Take photos of stained oil droplets using an inverted fluorescent microscope (FSX100, Olympus LS, Japan) (Fig. 12.4).

Notes

1. Add 2 mL gelatin for a 60-mm dish and 5 mL for a 100-mm dish
2. Trypsinization usually takes 2–3 min or less, do not incubate more for than 5 min
3. During the wash it should be kept in the dark.

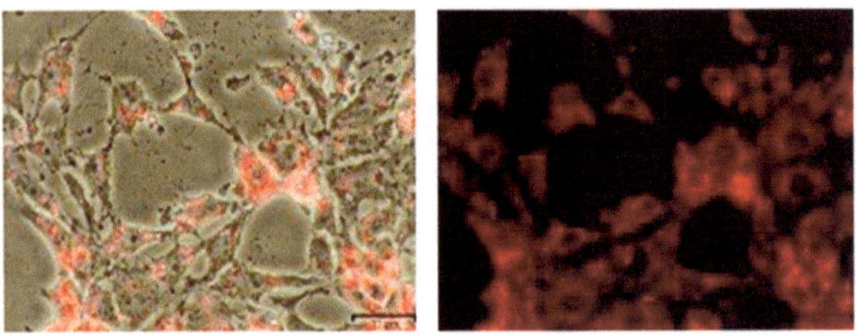

Fig. 12.4 Representative image of Nile Red staining for adipose droplets in miPS-MB231cm cells. Fluorescence (right) and merged with a bright field (left)

References

Afify SM, Hassan G, Osman A, Calle AS, Nawara HM, Zahra MH, El-Ghlban S, Mansour H, Alam MJ, Abu Quora HA, Du J, Seno A, Iwasaki Y, Seno M. Metastasis of cancer stem cells developed in the microenvironment of hepatocellular carcinoma. Bioengineering (Basel). 2019;6(3):73.

Afify SM, Hassan G, Seno A, Seno M. Cancer-inducing niche: the force of chronic inflammation. Br J Cancer. 2022a;127(2):193–201.

Afify SM, Hassan G, Yan T, Seno A, Seno M. Cancer stem cell initiation by tumor-derived extracellular vesicles. Methods Mol Biol. 2022b;2549:399–407.

Afify SM, Seno M. Conversion of stem cells to cancer stem cells: undercurrent of cancer initiation. Cancers (Basel). 2019;11(3):345.

Al-Hajj M, Wicha MS, Benito-Hernandez A, Morrison SJ, Clarke MF. Prospective identification of tumorigenic breast cancer cells. Proc Natl Acad Sci U S A. 2003;100:3983–8.

Chen L, Liu Y, Xu Y, Afify SM, Gao A, Du J, Liu B, Fu X, Liu Y, Yan T, Zhu Z, Seno M. Up-regulation of Dsg2 conferred stem cells with malignancy through wnt/β-catenin signaling pathway. Exp Cell Res. 2022;113416

Collins AT, Berry PA, Hyde C, Stower MJ, Maitland NJ. Prospective identification of tumorigenic prostate cancer stem cells. Cancer Res. 2005;65:10946–51.

Cozzo AJ, Fuller AM, Makowski L. Contribution of adipose tissue to development of cancer. Compr Physiol. 2017;8:237–82.

D'Esposito V, et al. Adipose microenvironment promotes triple negative breast cancer cell invasiveness and dissemination by producing CCL5. Oncotarget. 2016;7(17):24495–509.

Dirat B, et al. Cancer-associated adipocytes exhibit an activated phenotype and contribute to breast cancer invasion. Cancer Res. 2011;71(7):2455–65.

Eramo A, Lotti F, Sette G, Pilozzi E, Biffoni M, Di Virgilio A, Conticello C, Ruco L, Peschle C, De Maria R. Identification and expansion of the tumorigenic lung cancer stem cell population. Cell Death Differ. 2008;15:504–14.

Esper RJ, Nordaby RA, Vilariño JO, Paragano A, Cacharrón JL, Machado RA. Endothelial dysfunction: a comprehensive appraisal. Cardiovasc Diabetol. 2006;5:4.

Fujisaki K, et al. Cancer-mediated adipose reversion promotes cancer cell migration via IL-6 and MCP-1. Breast Cancer Res Treat. 2015;150(2):255–63.

Hassan G, Afify SM, Nair N, et al. Hematopoietic cells derived from cancer stem cells generated from mouse induced pluripotent stem cells. Cancers (Basel). 2019;12(1):82. Published 2019 Dec 29. https://doi.org/10.3390/cancers12010082.

Jain RK, Carmeliet P. SnapShot: tumor angiogenesis. Cell. 2012;149(1408–1408):e1.

Kumon K, Afify SM, Hassan G, Ueno S, Monzur S, Nawara HM, Quora HAA, Sheta M, Xu Y, Fu X, Zahra MH, Seno A, Seno M. Differentiation of cancer stem cells into erythroblasts in the presence of CoCl2. Sci Rep. 2021;11(1):23977.

Lee TK, Castilho A, Cheung VC, Tang KH, Ma S, Ng IO. CD24+ liver tumor-initiating cells drive self-renewal and tumor initiation through STAT3-mediated NANOG regulation. Cell Stem Cell. 2011;9:50–63.

Li C, Heidt DG, Dalerba P, Burant CF, Zhang L, Adsay V, Wicha M, Clarke MF, Simeone DM. Identification of pancreatic cancer stem cells. Cancer Res. 2007;67:1030–7.

Ma S, Chan KW, Hu L, Lee TK, Wo JY, Ng IO, Zheng BJ, Guan XY. Identification and characterization of tumorigenic liver cancer stem/progenitor cells. Gastroenterology. 2007;132:2542–56.

Mansour H, Hassan G, Afify SM, Yan T, Seno A, Seno M. Metastasis model of cancer stem cell-derived Tumors. Methods Protoc. 2020;3(3):60.

Mansour H, Afify SM, Hassan G, Abu HA, Quora HM, Nawara MH, Zahra JD, SadiaMonzur TO, Seno A, Seno M. A comparative study of metastatic potentials of three different cancer stem cell models. Adv Cancer Biol Metastasis. 2022;5:100062.

Michiels C. Endothelial cell functions. J Cell Physiol. 2003;196:430–43.

Nawara HM, Afify SM, Hassan G, Zahra MH, Atallah MN, Seno A, Seno M. An assay for cancer stem cell-induced angiogenesis on chick chorioallantoic membrane. Cell Biol Int. 2021;45(4): 749–56.

Osman A, Afify SM, Hassan G, Fu X, Seno A, Seno M. Revisiting cancer stem cells as the origin of cancer-associated cells in the tumor microenvironment: a hypothetical view from the potential of iPSCs. Cancers (Basel). 2020b;12(4):879.

Osman A, Oze M, Afify SM, Hassan G, El-Ghlban S, Nawara HM, Fu X, Zahra MH, Seno A, Winer I, Salomon DS, Seno M. Tumor-associated macrophages derived from cancer stem cells. Acta Histochem. 2020a;122(8):151628.

Prieto-Vila M, Yan T, Calle AS, et al. iPSC-derived cancer stem cells provide a model of tumor vasculature. Am J Cancer Res. 2016;6(9):1906–21. Published 2016 Sep 1

Takebe N, Miele L, Harris PJ, Jeong W, Bando H, Kahn M, Yang SX, Ivy SP. Targeting notch, hedgehog, and Wnt pathways in cancer stem cells: clinical update. Nat Rev Clin Oncol. 2015;12:445–64.

Wagenblast E, Soto M, Gutiérrez-Ángel S, Hartl CA, Gable AL, Maceli AR, Erard N, Williams AM, Kim SY, Dickopf S, Harrell JC, Smith AD, Perou CM, et al. A model of breast cancer heterogeneity reveals vascular mimicry as a driver of metastasis. Nature. 2015;520:358–62.

Weis SM, Cheresh DA. Tumor angiogenesis: molecular pathways and therapeutic targets. Nat Med. 2011;17:1359–70.

Yan T, et al. Characterization of cancer stem-like cells derived from mouse induced pluripotent stem cells transformed by tumor-derived extracellular vesicles. J Cancer. 2014;5(7):572–84.

Chapter 13
Tumor Angiogenesis by Cancer Stem Cells In Vivo

Abstract In response to a variety of cytokines, vascular endothelial cells proliferate and develop new tube structures from pre-existing blood vessels in quite an orderly manner during normal angiogenesis. However, abnormal angiogenesis contributes to the vast majority of diseases including cancer. Tumor angiogenesis plays a critical role in tumor development, progression, and metastasis.

Many reports evaluated that cancer stem cells are responsible for the secretion of angiogenic factors such as basic fibroblast growth factor (bFGF) and vascular endothelial growth factor (VEGF). Also, CSCs were reported to have the potential to differentiate into vascular endothelial cells developing neo-vasculature structures.

This chapter offers an overview of the role of CSCs in tumor angiogenesis assessed on chick chorioallantoic membrane (CAM). Furthermore, we show the detailed technique to design the model of CSC-based tumor angiogenesis and propose the possible application to target anti-angiogenesis. This evaluation employing CSCs should help develop more effective anti-angiogenic therapies in the future.

Keywords Angiogenesis · CAM · CSC · Stem cells

13.1 Introduction

Angiogenesis is a concept of vascularization developing new blood vessels from existing blood vessels (Ribatti and Crivellato 2012). The first scientific insight into angiogenesis was proposed by Scottish anatomist and surgeon John Hunter (1840). Recapitulating the hypothesis and investigation, the close relationship between tumor growth and angiogenesis was not realized until when Judah Folkman hypothesized that prevention of new blood vessel formation in small cancers could prevent the growth (Folkman 1971).

Tumor-induced angiogenesis appears to be initiated by angiogenic factors such as VEGF and FGF2, which stimulate the growth of vascular endothelial cells to induce neovascularization resulting in the tumor growth. Many scientists tried to demonstrate the link between CSCs and angiogenesis. Nowadays, it has become accepted

S. M. Afify, M. Seno, *Methods in Cancer Stem Cell Biology*,
https://doi.org/10.1007/978-981-99-1331-2_13

that CSCs significantly contribute to the tumor angiogenesis because CSCs have the capacity to differentiate into endothelial cell phenotype (Wang et al. 2010; Matsuda et al. 2014; Nawara et al. 2020; Nawara et al. 2021).

Blood supply can also follow vasculogenic mimicry (VM), which is independent of vascular endothelial cells. Although the link between CSC and tumor-induced angiogenesis or VM is still not fully understood, a few scientists demonstrated that CSCs are capable of transdifferentiating and forming vascular vessel-like structure in the absence of endothelial cells (Prieto-Vila et al. 2016).

In this context, the development of therapeutic agents that prevent the differentiation of CSCs into the cells with vascular phenotypes should be very critical for the development of effective therapy against cancer. To achieve this goal, comprehensive studies both in vitro and in vivo are needed for the modalities of novel anticancer therapy. Currently, the most commonly used models to evaluate the efficacy of drug candidates in vivo are mouse xenografts of human cancer cell lines (Jung 2014). On the other hand, researchers have employed chorioallantoic membrane (CAM)-assay as an effective in vivo model to study angiogenesis in a physiological and pathological context (Richardson and Singh 2003) since the use of CAM could provide an efficient alternative to animal models (Mapanao et al. 2021; Day et al. 2015). Further advantages in CAM are simple, efficient, cost effective, and easy to repeat, allowing cancer cell grafts (Tufan and Satiroglu-Tufan 2005).

CAM is employed in many reports to study tumor angiogenesis with tumor-derived cells and tumor microenvironment. In our method, we exploit CSCs developed from miPSCs (Afify et al. 2019, 2020) to evaluate tumor angiogenesis due to the capacity to differentiate into vascular endothelial cells. This differentiation process could be the novel point giving more reliable evaluation than ever because tumor cell grafts are expected not to differentiate but to induce the growth of vascular vessels of the host. Here, we provide a novel significance in the endothelial cell differentiation and a detailed development of the CAM model with using CSCs.

13.2 Materials

13.2.1 Reagents

- Liver CSCs (miPS-Huh7cm cells) (Afify, 2020) vail converted from mouse-induced pluripotent stem cells (miPSCs) (iPS-MEF-Ng-20D-17, Lot No. 012, Riken Cell Bank, Tokyo, Japan), in which puromycin (puro) resistant gene and green fluorescent protein (GFP) gene were cloned under the control of Nanog promoter.
- Corning® Matrigel®, growth factor reduced, phenol red-free (BD, cat. no. 356231).
- An egg-tray, for holding the eggs before culturing in the cups.
- Scissors, for cutting the eggshell after cracking.
- Rubber bands, for holding the cling wrap on the plastic cups.
- Toothpicks, for making pores.
- Forceps, for dropping the discs over the CAM.

- Fetal bovine serum (FBS, Gibco, Life Technologies, Massachusetts, USA, Cat. No: 10437-028).
- Dulbecco's modified Eagle's medium-high glucose (Wako, Osaka, Japan (catalogue number: 044-29,765)).
- Trypsin-EDTA (0.25%) (Nacalai Tesque, Kyoto, Japan, Cat. No: 327777-44).
- Penicillin/streptomycin mixed solution (100 U/mL) Nacalai Tesque, Kyoto, Japan, Cat. No-26253-84).
- 70% ethanol (Sigma-Aldrich; Cat. No.: 459836-2).
- Hank's balanced salt solution (HBSS) Genesee Scientific, El Cajon, USA.
- Liquid N2 storage tank.
- Phosphate buffered saline (PBS) (Genesee Scientific, El Cajon, CA, USA; Cat. no.: 25-508).

13.2.2 Equipment

- Eppendorf Centrifuge 5415R, Eppendorf AG, 22331 Hamburg, Germany.
- Sanyo MCO-19AIC(UV) CO_2 Incubator, Marshall Scientific, Hampton, USA.
- Type A2 Biological Safety Cabinets (E-Series).
- Olympus IX81 microscope (Olympus, Tokyo, Japan).
- Tissue culture-treated plate, 60 mm dish; TPP, AG Schaffhausen, Switzerland, Cat. No 93060.
- Tissue culture-treated plate,100 mm dish; TPP, AG Schaffhausen, Switzerland, Cat. No 93100.
- Filter max 250 mL, TPP, Switzerland, Cat. No 99255.
- Falcon® Conical Centrifuge Tubes (15 mL; BD Falcon, New York, USA, Cat. No 352095).
- Falcon® Conical Centrifuge Tubes (50 mL; BD Falcon, New York, USA, Cat. No 352070).
- 37 °C water bath.

13.2.3 Reagent Preparation

13.2.3.1 Stem Cell Medium

Mix the following for 500 mL		
Dulbecco's modified Eagle's medium-high glucose		412.5 mL
FBS		75 mL
Penicillin/streptomycin mixed solution	(50 U/mL)	2.5 mL
Gibco™ L-Glutamine(200 mM)	(2 mM)	5 mL
MEM non-essential amino acids solution	(1 mM)	5 mL

13.3 Methods

As a tool to investigate angiogenesis in vivo, we describe a unique method for CAM assays in chicks. CSC can be used for a highly efficient angiogenesis assay when combined with basic laboratory facilities and low-cost materials. By introducing CSC into this method, researchers will be able to evaluate issues related to angiogenesis that have been uncovered.

The CAM assay steps are so sensitive that much careful attention should seriously be paid. If you cannot start the experiment on the same day when you receive the fertilized eggs, they should be kept at 12 °C for 1 week. Although this is the recommended method to delay the development of fertilized eggs, the embryos should not be kept surviving under this harmful condition beyond 1 week. As we handled the experiment, we realized that it required a high level of concentration and patience. The most important factor leading to higher embryonic mortality was the creation of a window within the eggshell without disrupting the embryonic membranes. Figure 13.1 displays a scheme of the steps involved in CAM assays with additional details in the sections that follow.

13.3.1 Reviving CSC

1. Transfer 0.2% gelatin solution from 4 °C to the clean bench.
2. Cover the bottom of a 60-mm dish with sufficient volume of 0.2% gelatin solution.
3. Incubate the dishes for 30 min at 37 °C in an incubator with 5% CO_2.
4. Pick up vials of miPS-Huh7cm cells from the liquid nitrogen storage tank.
5. Thaw the cells quickly at 37 °C in a water bath.
6. Wipe the outside of the vial with 70% ethanol.
7. Add 5 mL of pre-warmed stem cell medium to a sterile 15-mL conical tube.
8. Suspend the cells gently with a sterile long Pasteur pipette.
9. Transfer the cells into the 15-mL conical tube containing the medium.
10. Centrifuge cells at 200 × g for 5 min at 25 °C.
11. Aspirate the supernatant without disturbing the pellet.
12. Add 5 mL of pre-warmed stem cell medium.
13. Resuspend the pellet by gently pipetting up and down 2 or 3 times with a sterile long Pasteur pipette.
14. Pick up the gelatin-coated dishes prepared in the previous step from the incubator.
15. Aspirate the gelatin solution at the bottom.
16. Add 4 mL of iPS medium.
17. Seed the cells on two gelatin-coated dishes dividing the cells into high and low numbers.
18. Remove the dead cells by changing medium on the next day.

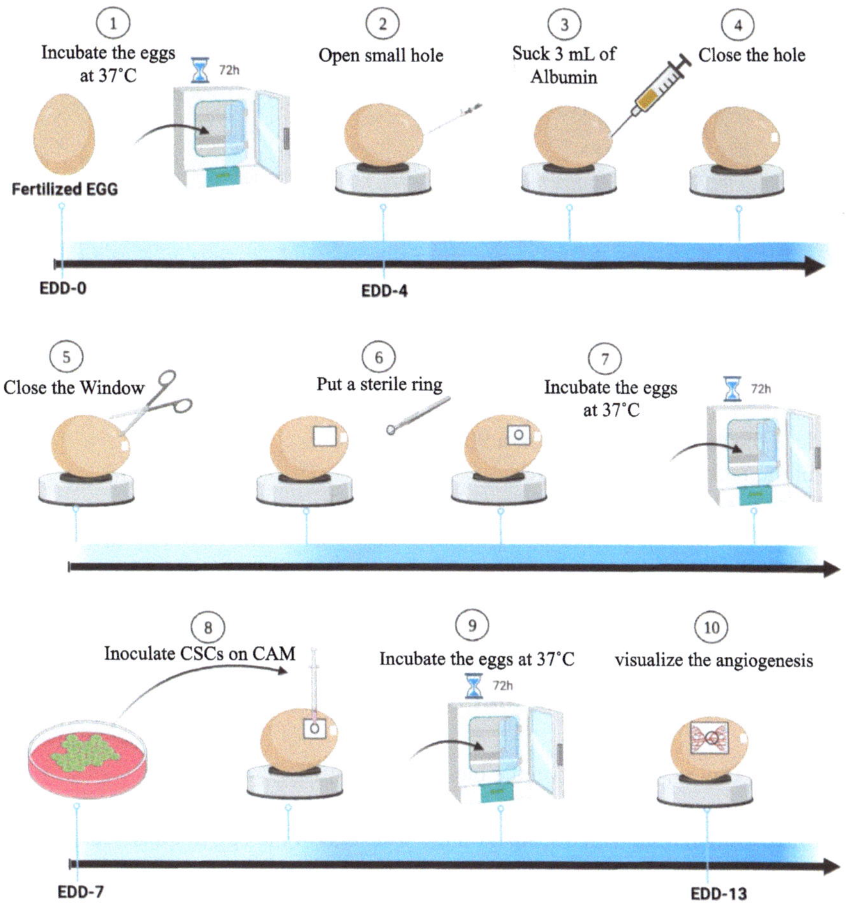

Fig. 13.1 A schematic representation of the CAM assay. After 4 days of incubation, 3 ml of albumin is removed, and a window is cut in the eggshell. At day 7 of incubation when the membrane is visible but not yet completely developed, the CSCs are implanted into the membrane. On day 13 of incubation angiogenesis was investigated. *CAM* chorioallantoic membrane, *CSC* cancer stem cells

19. Observe the cells and take photos every day (Fig. 13.2).
20. After three passages from reviving, the cells are ready for further experiments.

13.3.2 Incubation of the Fertilized Eggs (EDD-0)

1. Clean the shell of fertilized chicken eggs with distilled water.
2. Clean again with a piece of tissue paper soaked in 70% ethanol.

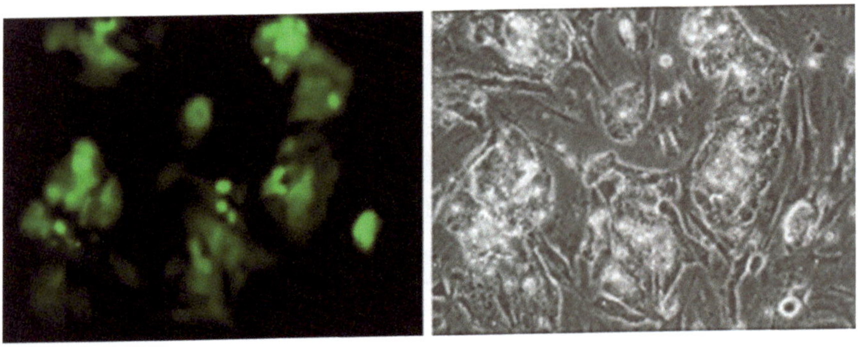

Fig. 13.2 Representative image for miPS-Huh7cmPcells. Lens (20X)

3. Place the eggs horizontally on a 90° in the rack.
4. Put the rack in the incubator.
5. Turn the rotation approximately 6 times per 24 h.
6. Incubate the fertilized chicken eggs at a humidity of 70% and a temperature of 37.8 °C.
7. Consider the first day of incubation as day 0 (EDD-0).

13.3.3 Puncture the Egg and Albumin Removal

1. Put the tray with the egg on the clean bench on EDD-4.
2. Place the eggs in an upright position.
3. Make a small hole with a scalpel on the wide end of the egg to translocate the air compartment in the egg to the top.
4. Re-organize the eggs horizontally in the tray.
5. Wait for 5 min.
6. Insert the sterile needle with a syringe into the hole at an angle of approximately 45° to a depth of approximately 1–1.25 cm (Fig. 13.3). Note 1.
7. Suck 3–4 ml of egg white.
8. Cover the hole with a small piece of sterile tape.
9. Return the egg to its original position.
10. Incubate for 1 h at 37 °C adjusting the humidity 50% without rotation.
11. Place the eggs back in the incubator without rotation. Note 2.
12. Adjust the temperature to 37.8 °C and the humidity to 70% in the incubator.
13. Incubate the eggs for 3 days (EDD-6).

13.3.4 Opening a Window in the Egg over CAM

1. Take out the eggs from the incubator.
2. Open a window of approximately 1 cm^2 at the top of the shell with tweezers cutting off the tape from each egg.

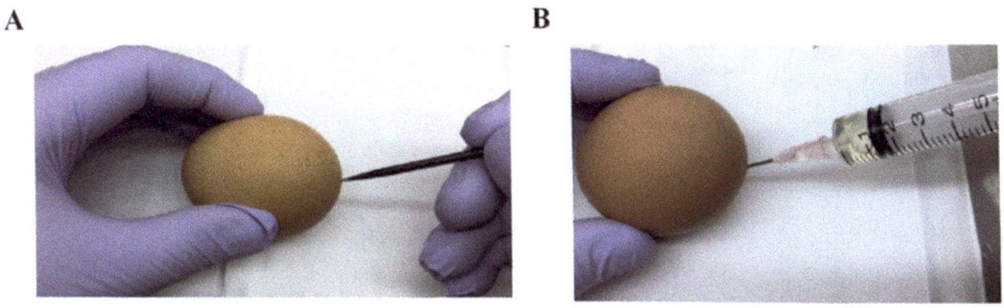

Fig. 13.3 At EDD-4 of incubation, a small hole is punched into the eggshell at the "bottom" of the egg (**a**). Then, around 3 ml of egg white is sucked with a syringe and the hole is closed with adhesive tape in a sterile condition (**b**)

3. Check embryos in the eggs.
4. Find live eggs.
5. Discard dying embryos.
6. Remove the tape from the egg under a stereoscopic microscope.
7. Observe and photograph the CAM vasculature through the window. Note 3.
8. Put a sterile ring on the CAM (Fig. 13.4). Note 4.
9. Close the window and incubate the egg until cells become ready.

Finally, the window is closed with an additional stripe of tape

13.3.5 *Inoculation of CSCs on CAM*

1. Pick up the CSCs in a dish from the incubator immediately after the embryos become ready for inoculation.
2. Wash the cells with 2 mL of PBS.
3. Repeat washing twice to remove dead cells.
4. Add 2 mL of 0.05% trypsin and incubate for 5 min at 37 °C until the cells detach.
5. Stop trypsinization by adding 8 mL of fresh complete culture media.
6. Transfer the cells to a 15-mL sterile falcon tube.
7. Centrifuge cells at 200 × g for 5 min at 25 °C.
8. Aspirate the supernatant without disturbing the pellet.
9. Add 5 mL of pre-warmed stem cell medium.
10. Count the cells with a hemocytometer or an automated cell counter.
11. Dilute cells to 1000 cells/40 μL with serum-free medium and containing Matrigel in 50%.
12. Bring the egg with a window opened in the previous step.
13. Inoculate 50 μL of the cell suspension into the ring set on CAM.
14. Reseal the window with sterile laboratory tape.
15. Return the eggs to the incubator.

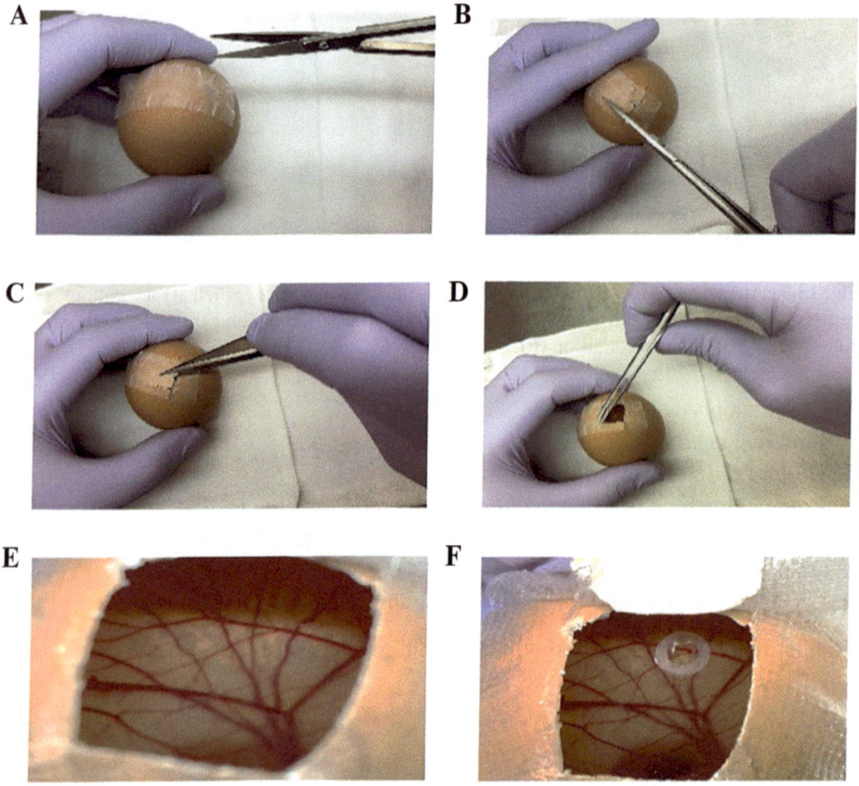

Fig. 13.4 Opening a window in the eggs. A strip of tape is glued on the side of the egg (**a**). A window is cut in the area of the scotch to avoid eggshell crumbs to fall on the membrane, using sharp scissors (**b–d**). Then a sterile ring was inserted on the CAM (**e**, **f**)

13.3.6 Observation of Tumor Angiogenesis

1. Place the egg on the egg holder and remove the tape covering the window on day 13 (EDD-13).
2. Widen the window carefully removing the shell to increase the field of view without disturbing the extra-embryonic membrane or embryo.
3. Inject approximately 2–3 ml of emulsified fat warmed to 37 °C under the CAM until the whole field becomes white to increase contrast for imaging and to obstruct the underlying vessels of the vitelline membrane (Fig. 13.5).
4. Capture the vascular patterns with a digital camera at high-resolution.
5. Quantify the blood vessels on the 13th day from the photograph. (Note 5).
6. Compare the results with other groups including negative/positive controls.
7. Take the membrane and perform immunofluorescence analysis according to the manufacturer's protocol using Anti-CD31 antibody (ab28364) as primary antibody, then incubated for 1 h at room temperature in the dark with rabbit

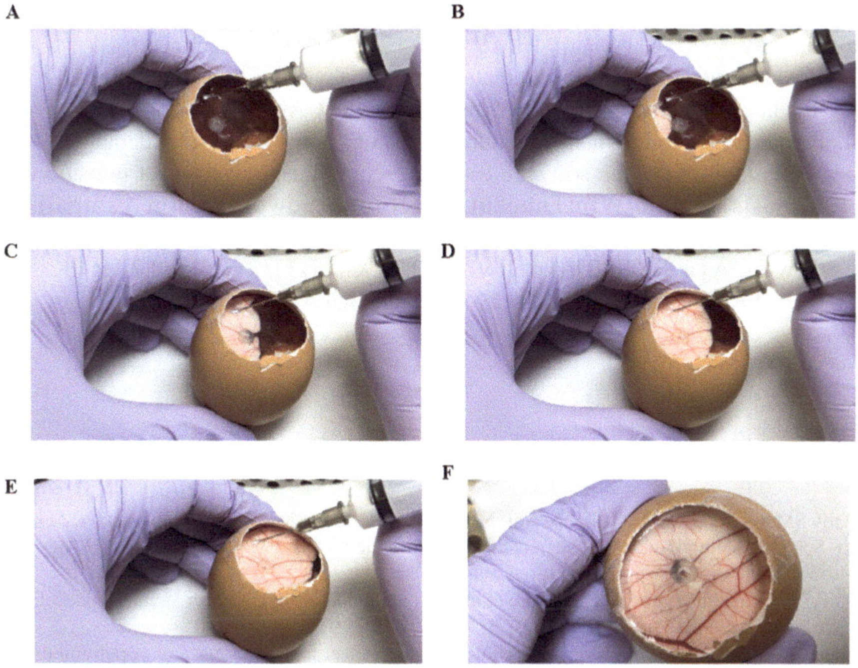

Fig. 13.5 Visualization of angiogenesis. During this time, the windows are enlarged to enable a more precise handling (**a**). If desired, pictures of the tumors in the egg can now be taken after injecting approximately 2–3 ml of emulsified fat (**b**–**f**)

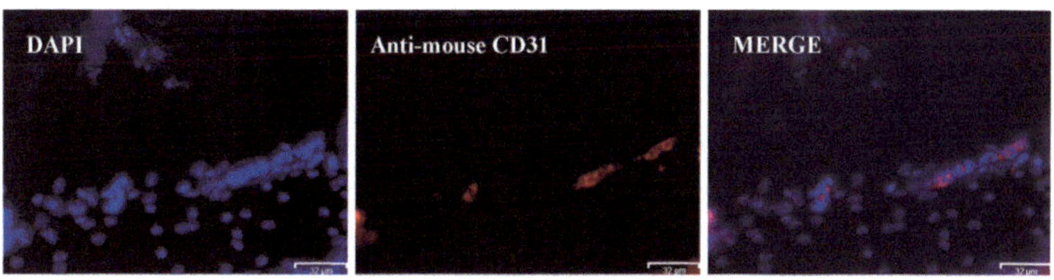

Fig. 13.6 Representative images of CD31 expression (middle right) and DAPI (left) and merge of immunoreactive CD31 and DAPI (Right)

immunoglobulin G (H + L) cross-adsorbed secondary antibody (A-21428) (Fig. 13.6).

Notes

1. Be careful not to puncture the CAM underneath when making the hole in the eggshell membrane. If bleeding occurs, the embryo will not survive.

2. Make the previous steps as quickly as possible in less than 40 min in order to keep viability.
3. If you cannot observe the vasculature, close the window and wait until EDD-7 or -8 when you can confirm the vasculature on the CAM.
4. Prepare rings cutting a silicone tube (i.d. 3 or 4 mm) with 2-mm depth.
5. Wimasis provides the software for microscopic image analysis at https://www.wimasis.com/en/WimTube?utm_medium=adwords&utm_campaign=SEARCH&utm_source=tube_formation_assay&gclid=CjwKCAjw8sCRBhA6EiwA6_IF4X4eizrItaml7gp17dm1Kym_QpPXPRyRnk11WKHTXSOx2S-9R9wmphoCpA8QAvD_BwE.

References

Afify SM, Chen L, Yan T, Calle AS, Nair N, Murakami C, Zahra MH, Okada N, Iwasaki Y, Seno A, Seno M. Method to convert stem cells into cancer stem cells. Methods Protoc. 2019 Aug 16;2(3):71.

Afify SM, Sanchez Calle A, Hassan G, et al. A novel model of liver cancer stem cells developed from induced pluripotent stem cells. Br J Cancer. 2020;122(9):1378–90.

Day C-P, Merlino G, Van Dyke T. Preclinical mouse cancer models: a maze of opportunities and challenges. Cell. 2015;163(1):39–53.

Folkman J. Tumor angiogenesis: therapeutic implications. N Engl J Med. 1971;285(21):1182–6. https://doi.org/10.1056/NEJM197111182852108.

Hunter J. A treatise on the blood, inflammation and gunshot wounds. Palmer JF (ed). p. 195, 1794. Philadelphia: Raswell, Barrington, and Haswell; 1840.

Jung J. Human tumor xenograft models for preclinical assessment of anticancer drug development. Toxicol Res. 2014;30(1):1–5.

Mapanao AK, Che PP, Sarogni P, Sminia P, Giovannetti E, Voliani V. Tumor grafted – Chick Chorioallantoic membrane as an alternative model for biological cancer research and conventional/nanomaterial-based Theranostics evaluation. Expert Opin Drug Metab Toxicol. 2021;17(8):947–68.

Matsuda S, Yan T, Mizutani A, Sota T, Hiramoto Y, Prieto-Vila M, Chen L, Satoh A, Kudoh T, Kasai T, Murakami H, Fu L, Salomon DS, Seno M. Cancer stem cells maintain a hierarchy of differentiation by creating their niche. Int J Cancer. 2014;135(1):27–36.

Nawara HM, Afify S, Hassan G, et al. Paclitaxel and sorafenib: the effective combination of suppressing the self-renewal of cancer stem cells. Cancers (Basel). 2020;12(6):1360. Published 2020 May 26

Nawara HM, Afify SM, Hassan G, Zahra MH, Atallah MN, Seno A, Seno M. An assay for cancer stem cell-induced angiogenesis on chick chorioallantoic membrane. Cell Biol Int. 2021;45(4):749–56.

Prieto-Vila M, Yan T, Calle AS, Nair N, Hurley L, Kasai T, Kakuta H, Masuda J, Murakami H, Mizutani A, Seno M. iPSC-derived cancer stem cells provide a model of tumor vasculature. Am J Cancer Res. 2016;6(9):1906–21.

Ribatti D, Crivellato E. "Sprouting angiogenesis", a reappraisal. Dev Biol. 2012;372(2):157–65.

Richardson M, Singh G. Observations on the use of the avian chorioallantoic membrane (CAM) model in investigations into angiogenesis. Curr Drug Targets-Cardiovasc Hematol Disord. 2003;3:155–85.

Tufan AC, Satiroglu-Tufan NL. The chick embryo chorioallantoic membrane as a model system for the study of tumor angiogenesis, invasion and development of anti-angiogenic agents. Curr Cancer Drug Targets. 2005;5(4):249–66.

Wang R, Chadalavada K, Wilshire J, et al. Glioblastoma stem-like cells give rise to tumour endothelium. Nature. 2010;468:829–33.

Chapter 14
Invasion and Metastatic Potential of Cancer Stem Cells In Vitro

Abstract In order to fight against cancer, it is critical to study cellular migration, invasion, and adhesion, which are the pivotal steps of metastasis. Characterization of the molecular cascade responsible for metastasis is still too complicated for analyses by the methods currently available. High-throughput screening in vitro to discover the small molecules against metastasis does not appear sophisticated. This is why most of the studies on cell migration are limited to be based on cancer cell line but not based on cancer stem cells. If cancer stem cells are available at hand, the metastatic events of cancer will be more reasonably explained. Therefore, the conventional methods to develop cancer stem cell from normal stem cells will be demanded. In this context, the method to convert induced pluripotent stem cells into cancer stem cells in the presence of conditioned medium of cancer cells will be a tremendously huge technique to understand cell migration and invasion in the future cancer research. Here in the current chapter, we present in vitro protocols that explain cell migration, invasion, and adhesion approaches step by step. Typically, scratch/wound healing assay and trans well chamber assay are introduced.

Keywords Invasion · Metastasis · Migration · Colongenic · Cancer stem cell

14.1 Introduction

As far as metastasis is the major cause of cancer death, the molecular mechanisms and cellular processes underlying metastasis will continue to be a major focus of cancer research. Metastasis is the spreading of tumor cells to other sites. This is a cascade of cellular invasion involving the alternative change of phenotypes termed "epithelial-mesenchymal transition (EMT)" and "mesenchymal-epithelial transition (MET)" (Fig. 14.1). The initial step of the metastatic cascade is local migration of the cells from the primary tumor site invading into the surrounding extracellular matrix (ECM). This process is initiated by the activation of signaling pathways that regulate adhesion, cytoskeletal dynamics, and movement within the tumor tissue (Friedl and Alexander 2011). This step involves breakdown of the basement membrane by enzymes such as proteases and glycanases releasing the tumor cells into the surrounding ECM. Then the tumor cells penetrate into blood circulation as the second

S. M. Afify, M. Seno, *Methods in Cancer Stem Cell Biology*,
https://doi.org/10.1007/978-981-99-1331-2_14

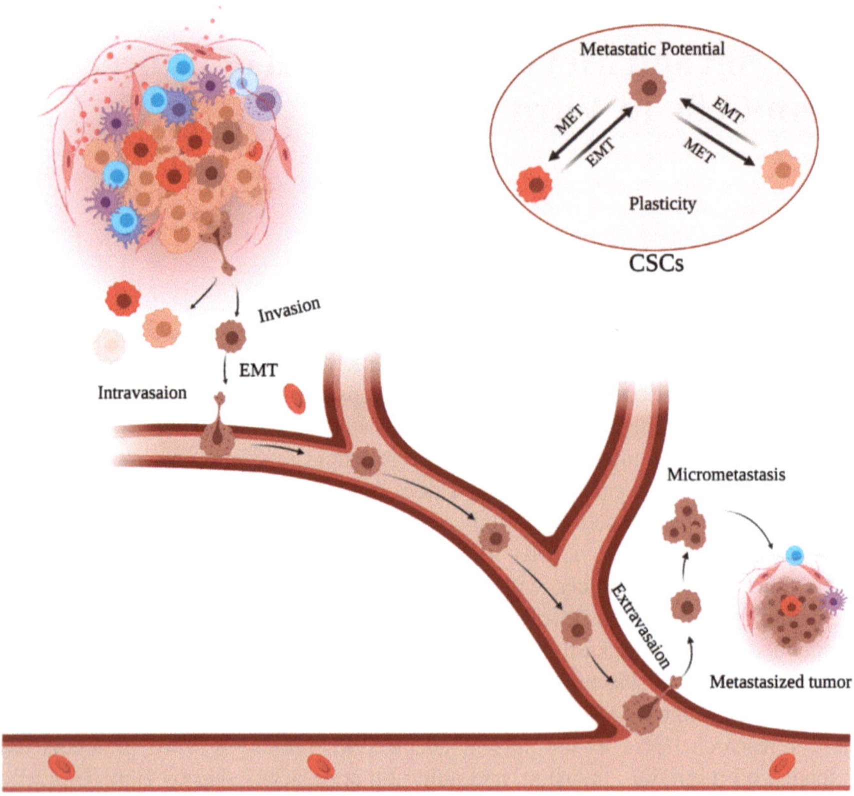

Fig. 14.1 The overview of metastasis. Tumor cells will metastasize to distal site in the body through the blood vessel. During the metastasis cancer stem cells play a significant role undergoing EMT and MET, which owe the character of plasticity without losing stemness. As a result, CSCs can localize in the primary tumor and in the metastasized tumors. These two different tumors could sometimes exhibit the same characters but sometimes different ones due to the plasticity of CSC being affected by the local microenvironment

step which is the so-called intravasation. After entering into the circulation, tumor cells can disseminate widely throughout the body along with the blood flow. They are known as circulating tumor cells (CTCs).

Surviving tumor cells can sometimes be intercepted in small capillaries or sometimes actively adhere with the molecular affinity to the vessel walls even in larger vessels and then extravasate into tissues and organs (Reymond et al. 2013). Once settled down at the site of metastasis, tumor cells are referred to as disseminated tumor cells (DTCs). DTCs are well known to be latent and quiescent for years or decades as small colonies (Massagué and Obenauf 2016), which often start to grow aggressively after the surgery of the primary tumors. The growth of micrometastases of DTCs are recognized as colonization.

In several studies, it has been revealed that tumors are highly heterogeneous and that only a small number of subpopulations can invade the basal membrane and metastasize to distant locations.

Increasing evidence demonstrated the presence of a very small population of CSCs which are associated with the aggressiveness characterized by increased ability to survive, resistance to treatment, and recurrence of tumors (Ajani et al. 2015, Afify and Seno 2019, Afify et al. 2022). Although the mechanism of metastasis is still unclear, CSCs are reported to be responsible for the metastasis (Afify et al. 2019, Mansour et al. 2020, 2022). Cancer stem cells (CSCs) contribute to metastasis through their stem-like plasticity, which enables them to initiate and complete the differentiation required for the spread of cancer (Dalerba et al. 2011). Some evidence suggested that the metastatic feature of CSC could be explained by the cell plasticity lying in the original CSCs that fluctuated between mesenchyme and epithelium throughout tumor progression (Fig. 14.1).

Although metastasis involves multiple processes, it is simply defined by the invasion in initial tissue and migration from the primary site in a tumor to the other. Therefore, it is essentially important to analyze the molecular mechanism of the process that cancer cells acquire an invasive phenotype.

Over recent years, scientists tried to develop innovative in vitro assays that mimic the metastatic cascade in part or in whole. However, most of the studies on cell migration and invasion have faced the limitation based on cancer cell lines. New approaches in vitro with new models of cancer cells and CSCs appear essential for further challenges. Various biological methods will preferably be employed to study these events in detail. The cell culture wound-closure and invasion assays are widely used in the scientific community (Castellone et al. 2011; Albini 1998). This chapter provides a brief overview of both assays of cell migration and invasion with the CSC model derived from iPSCs.

14.2 Materials

14.2.1 Reagents

- Cancer stem cells (e.g., miPS-Huh7cm cells) (Afify et al. 2020).
- Dulbecco's modified Eagle medium (DMEM, Sigma-Aldrich, St. Louis, MO, USA).
- Disodium hydrogen phosphate (Na_2HPO_4) (Sigma-Aldrich, cat. no. 255793).
- Fetal bovine serum (FBS, e.g., Gibco, Life Technologies, Massachusetts, USA) (Note 1).
- Penicillin/streptomycin mixed solution (Nacalai Tesque, Japan).
- Phosphate-buffered saline (Dulbecco's formula PBS) (Genesee Scientific, El Cajon, USA).
- Potassium chloride (KCl) (Sigma-Aldrich, cat. no. P9333).
- Potassium dihydrogen phosphate (KH_2PO_4) (Sigma-Aldrich, cat. no. P0662).

- MEM non-essential amino acids solution (×100) (Thermo Fisher Scientific, Waltham, MA, USA).
- L-Glutamine (Nacalai Tesque, Japan).
- Sodium chloride (NaCl) (Sigma-Aldrich, cat. no. 746398).
- 2-Mercaptoethanol (Sigma-Aldrich, St. Louis, USA).
- 2.25% Trypsin-EDTA (Atlanta Biologicals, Flowery Branch, USA).
- Gelatin, tissue culture grade (type B, 2% in H_2O; Sigma-Aldrich, cat. no. G1393).

14.2.2 Equipment

- Adjustable pipettes: P-20 (Gilson, cat. no. FA10003M), P-200 (Gilson, cat. no. FA10005M) and P-1000 (Gilson, cat. no. F123602M).
- Biosafety Cabinet (BSC-04IIA2) airtech.
- Safety Cabinet Class 2 (ESCO, model no. LA2-4A1).
- Cell culture incubator (CO_2 at 5%, humidified at 37 °C) (Thermo Scientific cat. no. 311).
- CKX41 Inverted Microscope – Olympus Life Science (Japan).
- Cotton swabs (Sigma-Aldrich, cat. no. Z699365).
- Eppendorf 5415R Refrigerated Centrifuge (Eppendorf, Germany).
- Falcon® Conical Centrifuge Tubes (15 mL; BD Falcon, New York, NY, USA, Cat. No 352095).
- Hemocytometer (Sigma-Aldrich, cat. no. Z359629).
- Laser scanning confocal microscope, FV-1000, (Olympus, Tokyo, Japan).
- Liquid N2 storage tank.
- Microcentrifuge 1.5-mL tubes (Eppendorf, cat. no. 0030120086).
- Pipette tips 20 µl (Starlab, cat. no. S1120–1810), Pipette tips: 200 µl (Starlab, cat. no. S1120–8810), Pipette tips 1000 µl (Starlab, cat. no. S1120–1830).
- Sterile forceps.
- Syringes.
- Software/ImageJ – Computer Vision Online.
- Water bath (Thermo Scientific cat. no. 152–4101).
- Tissue culture-treated plate, 60 mm dish (TPP Techno Plastic Products AG Schaffhausen, Switzerland, Cat. No 93060).
- Tissue culture-treated plate, 100 mm dish (TPP Techno Plastic Products AG Schaffhausen, Switzerland, Cat. No 93100).
- Transwell inserts for 24-well plates (membrane 8.0 µm pores) (Corning, cat. no. 353097).
- 5/10/25-mL plastic disposable pipette.

14.2.3 *Reagent Preparation*

14.2.3.1 Medium for Cancer Stem Cells Culture

For cancer stem cells, prepare high glucose containing DMEM supplemented with final concentrations of 15% FBS, 0.1 mM MEM non-essential amino acids solution, 2 mM L-glutamine, 0.1 mM 2-mercaptoethanol, 50 U/mL penicillin/streptomycin. Store at 4 °C (Note 2).

14.2.3.2 Matrigel Preparation

Thaw aliquots of Matrigel slowly at 4 °C overnight. Mix 100 μl of Matrigel and 300 μl of cold DMEM (without FBS) in a sterile 1.5-ml microcentrifuge tube (Note 3).

14.2.3.3 PBS (pH 7.40) Preparation

Prepare PBS by mixing 137 mM NaCl, 10 mM Na_2HPO_4, 2.7 mM KCl, and 1.8 mM KH_2PO_4. Prepare the buffer in distilled water (dH_2O), adjust the pH to 7.4 and autoclave it for 20 min (Note 4).

14.2.3.4 0.1%(W/V) Gelatin

Dissolve 0.5 g of gelatin (from porcine skin) in 500 ml distilled water and autoclave. Store at 4 °C.

14.3 Methods

All the following steps must be performed in a standard tissue culture hood in an aseptic way. Contamination and overgrowth should be avoided all through the experiment. Cells should be passaged at least 3 times before starting the experiment.

14.3.1 *Preparation of Gelatin-Coated Dishes*

1. Transfer 0.2% gelatin solution from 4 °C to the clean bench.
2. Cover the bottom of the dish with sufficient volume of 0.2% gelatin solution.
3. Incubate the dishes for 30 min at 37 °C in an incubator with 5% CO_2.

14.3.2 Cancer Stem Cell Thawing

1. Coat three 60-mm dishes with 0.1% gelatin at 37 °C in an incubator balanced with 5% CO_2 for at least 30 mins.
2. Pre-warm cancer stem cell culture medium at 37 °C.
3. Transfer 5 ml of pre-warmed culture medium to a sterile 15-ml conical tube.
4. Pick up miPS-Huh7cm cells from liq N_2.
5. Thaw in a 37 °C water bath incubating around 1–3 mins.
6. Take the vial of cells inside of the cell culture hood.
7. Sterilize the vial wiping with cotton in 70% ethanol inside of the cell culture hood.
8. Transfer the cells into the 15-ml conical.
9. Add 5 ml of pre-warmed medium slowly into the tube dropping one by one with gentle mixing.
10. Centrifuge at 200 × g for 5 mins.
11. Remove the supernatant without disturbing the cell pellet.
12. Resuspend the cell pellet in 1 mL of pre-warmed medium.
13. Seed the cells on gelatin-coated dishes (Note 5).
14. Incubate the cells at 37 °C, in an incubator balanced with 5% CO_2.
15. Try to make medium changes with 2-day intervals observing the cell conditions every day (Note 6).
16. Take photos if necessary with fluorescence microscope, FV-1000, Olympus, Tokyo, Japan (Fig. 14.2).

14.3.3 Cancer Stem Cell Passage

The cell growth rate will get stable in approximately 1 week after thawing. Passage number should be recorded for each passage. Cells are expected to be ready for evaluation of migration and invasion potential after three passages. Start the following procedure when the cell density reaches 70% confluent in a 60-mm dish.

1. Aspirate the medium in the culture dish.
2. Rinse the cells in the dish twice with sterile PBS.
3. Add 2 mL of 0.05% trypsin to the dish.
4. Incubate the dish for 5 min at 37 °C.
5. Stop trypsinization by adding FBS containing pre-warmed medium.
6. Harvest the cells and transfer to a 15-ml conical tube.
7. Spin down the cells at 200 g for 5 min at 4 °C.
8. Aspirate the supernatant without disturbing the cell pellet.
9. Resuspend the pellet in 1 mL medium.
10. Determine the cell concentration with a hemocytometer.
11. Seed the cells at 0.3×10^6 on a gelatin-coated dish.

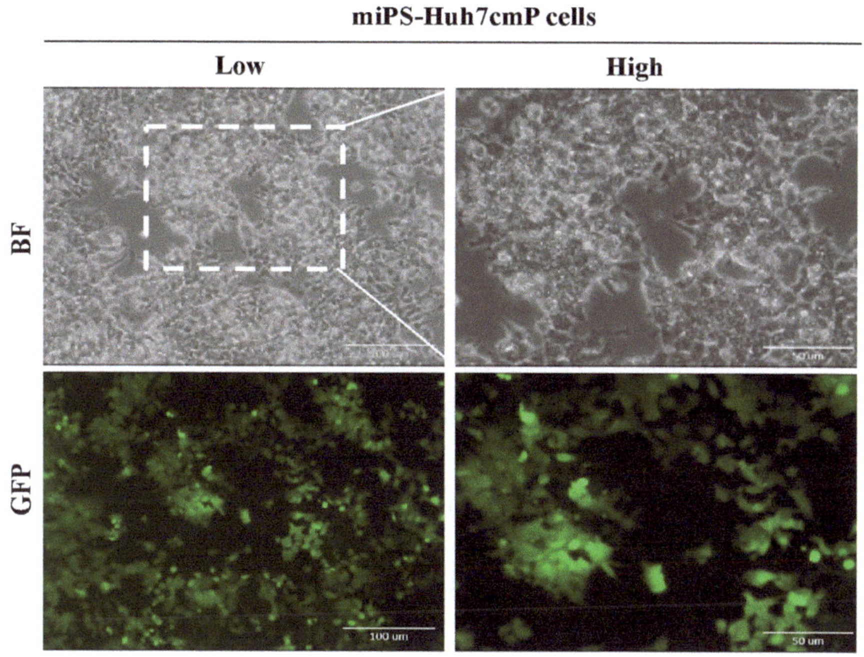

Fig. 14.2 Representative images for cancer stem cells (miPS-Huh7cm cells) derived from miPS cells in the presence of conditioned medium secreted from hepatocellular carcinoma. The green fluorescent protein (GFP) under the control of Nanog promoter allowing us to distinguish self-renewing undifferentiated CSCs from differentiated cells by the presence or absence of GFP expression. Low:×10 objective lens. High:×20 objective lens

12. Incubate the dish at 37 °C in a 5% CO_2 incubator.
13. Repeat Steps 1–11 for more than two passages.
14. After the third passage, cells are ready for further experiments.

14.3.4 Cell Migration

Cell migration can be monitored by the in vitro scratch assay based on the observation. When the monolayer of the cells covers the bottom of the dish surface up to approximately 85–90%, the experiments could start by creating an artificial gap, a so-called "scratch," on a confluent cell monolayer. The movement of the cells will be observed toward the gap space to close the "scratch" so that the new cell-to-cell contacts could be established again. As per the experimental procedure summarized in Fig. 14.3, a scratch is made in a confluent monolayer of miPS-Huh7cm cells using a pipette tip. After the debris is washed with PBS, FBS-free medium is added and cells are incubated for 24 h. Migration is evaluated by measuring the narrowing scratch by taking microscopic photos at different time points.

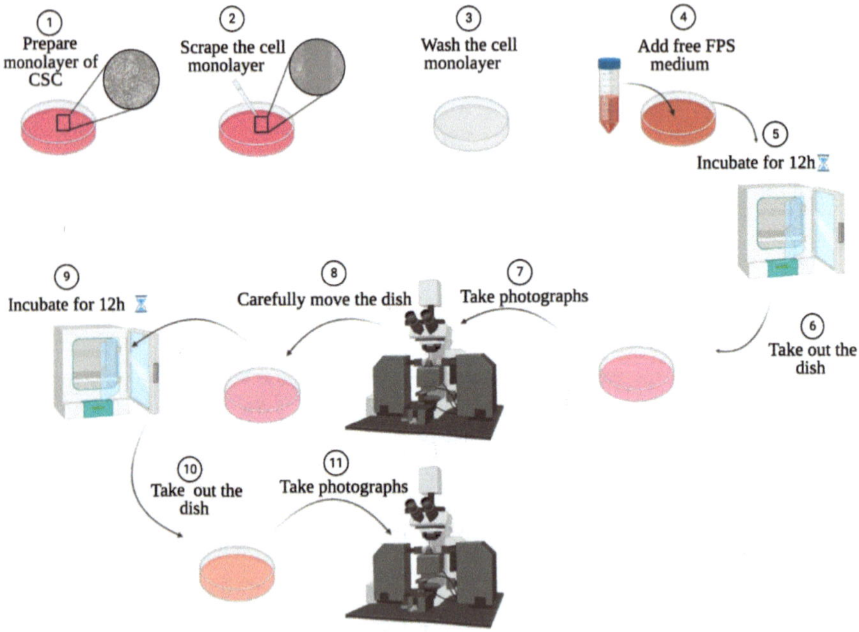

Fig. 14.3 Representative scheme for scratch assay procedure

1. After the third cell passage, incubate the cells at 37 °C in a 5% CO_2 incubator for 6–12 h allowing cells to adhere.
2. Monitor the cells under a microscope to confirm the monolayer of cell density becomes confluent.
3. Scrape the cell monolayer in a straight line to create a "scratch" with a P200 Gilson pipette tip.
4. Mark on the bottom of the dish with a marker pen to be sure about the direction of the scratch so that you can confirm it as a reference of the photographs.
5. Rinse the cells twice with 1 ml of sterile PBS to remove the floating cells and debris by careful aspiration.
6. Add 5 mL of low serum or serum-free DMEM.
7. Place the dish in an incubator at 37 °C in 5% CO_2 for 24 h.
8. Take photographs of the scratch periodically under a phase-contrast microscope and return the dish to the incubator (Fig. 14.4).
9. Analyze the open area of scratch from the photos by Wimasis software (Fig. 14.5).

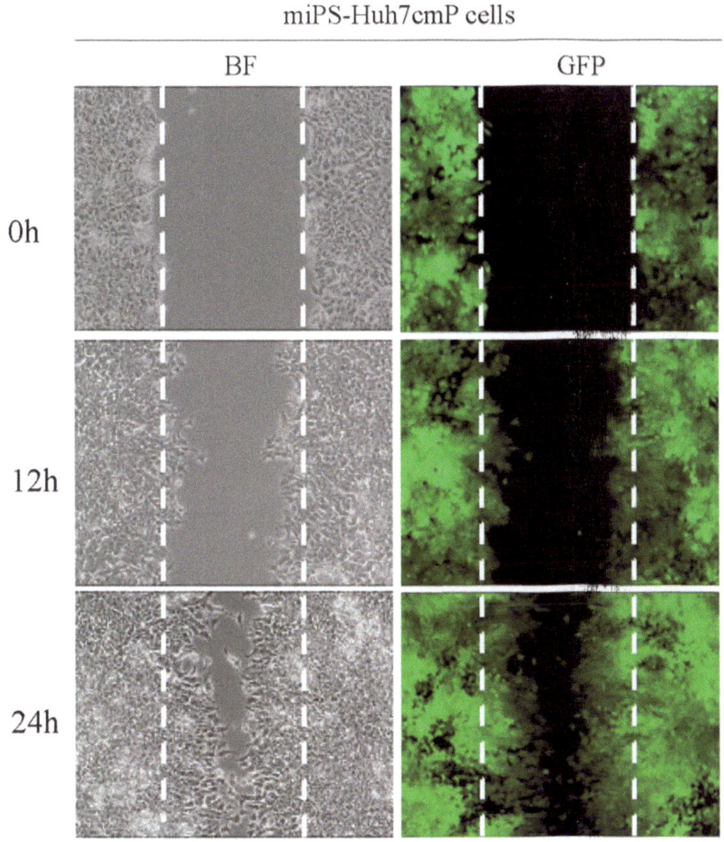

Fig. 14.4 Representative figure showed in vitro scratch assay for miPS-Huh7cm cells to track migration of individual cells in the edge of the scratch after 24 h of incubation. BF, bright field; GFP, green fluorescent protein

14.4 Cell Invasion

Invasion assay is used to measure the ability of cells to degrade the basement membrane. Transwell, a permeable filter insert, is available for this assay to measure the potential of invasion. This filter insert is positioned in a well of a culture plate and placed in direct contact to a medium supplemented with chemo-attractants such as serum or soluble growth factors. During this procedure of assay, any pipettes, syringes, or containers that will come in contact with the Matrigel matrix must be chilled prior to use. Before starting the experiment, plan the duration of the period in which cancer stem cells are allowed to invade. After the designated period, the cells that have invaded to the other side of the filter inserts should be fixed and stained, and then the extent of invasion should be quantified Fig. 14.6.

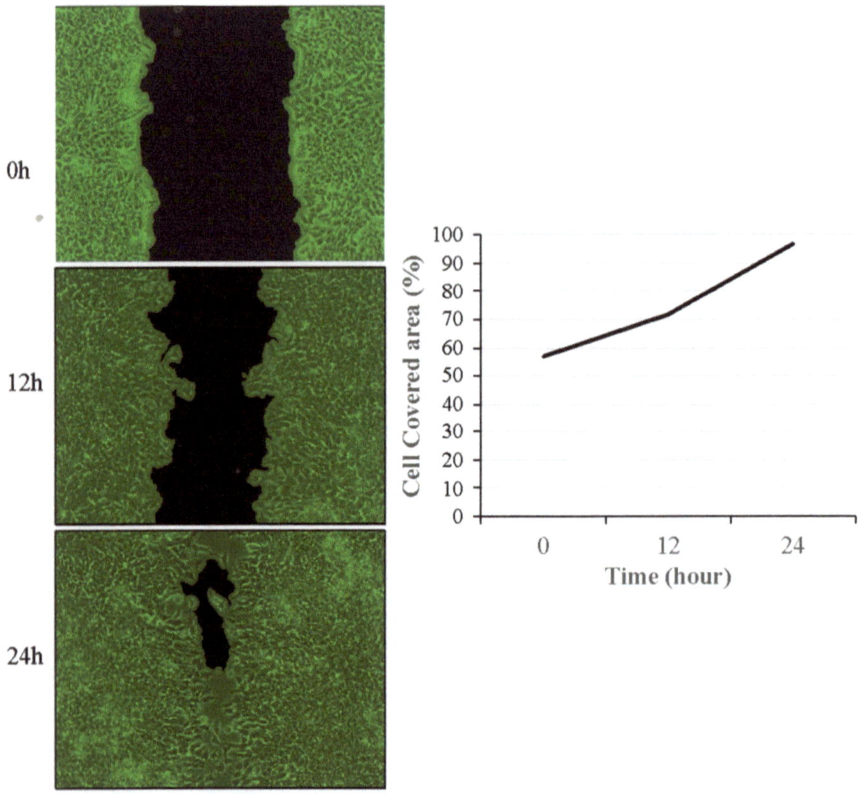

Fig. 14.5 Representative photos of miPS-Huh7cm cells in the scratch assay (left) and the Wimasis analysis from the photos during 24 h (right)

1. Incubate the CSC dish at 37 °C in a 5% CO_2 incubator after the third passage (Note 7).
2. Thaw Matrigel overnight at 4 °C prior to the experiment.
3. Put the insert in a 24-well plate and place at 4 °C.
4. Dilute 50 μl of Matrigel in 1 ml of cold serum-free DMEM on ice.
5. Add 100 μl of diluted Matrigel into the upper compartment of the Transwell insert.
6. Incubate the plate with the insert at 37 °C for 2 h until the Matrigel solidifies.
7. Pick up the CSC dish and aspirate the medium.
8. Wash the cells with sterile PBS.
9. Add 1 mL of 0.05% trypsin to the dish.
10. Incubate the dish for 5 min at 37 °C in a 5% CO_2 incubator.
11. Stop trypsinization by adding 2 mL pre-warmed medium supplemented with FBS.
12. Transfer the cell suspension to a sterile 15-mL centrifuge tube.
13. Spin down cells at 200 g for 5 min at room temperature.

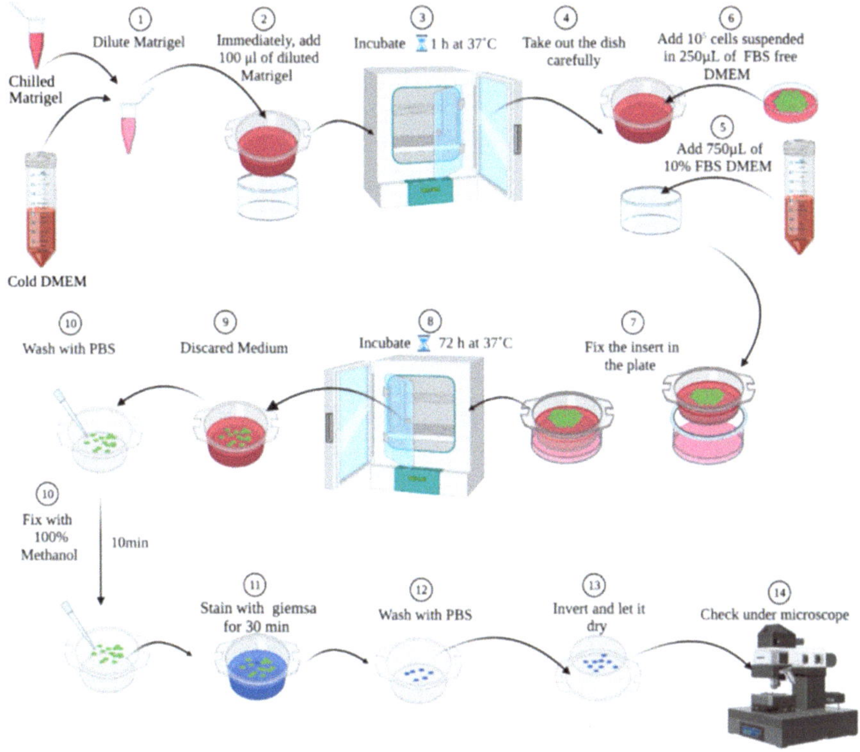

Fig. 14.6 Representative schemes for Transwell assay procedure using CSCs

14. Carefully aspirate the medium without disturbing the pellet.
15. Resuspend the cell pellet in 1 mL FBS-free DMEM.
16. Pick up 20 μl of cell suspension and mix with 20 μL of 0.5% trypan blue and determine the cell concentration with a hemocytometer.
17. Adjust the cell concentration to 10^5 per 250 μl in a sterile 1.5-mL tube (Note 8).
18. Remove the remaining medium from the permeable support membrane without disturbing the layer of Matrigel matrix on the membrane after Matrigel solidification.
19. Wash the insert twice with sterile PBS.
20. Add 750uL of DMEM supplemented with 10% FBS as attractant to the lower compartment of a well of the plate.
21. Put the insert into the well so the bottom membrane merges to the medium.
22. Gently transfer the cells from Step 17 to the upper compartment of the insert.
23. Incubate the 24-well plate with the inserts for 72 hours at 37 °C in a 5% CO_2 incubator.
24. Take out the plate carefully after the scheduled incubation period.
25. Wash the insert three times with PBS.

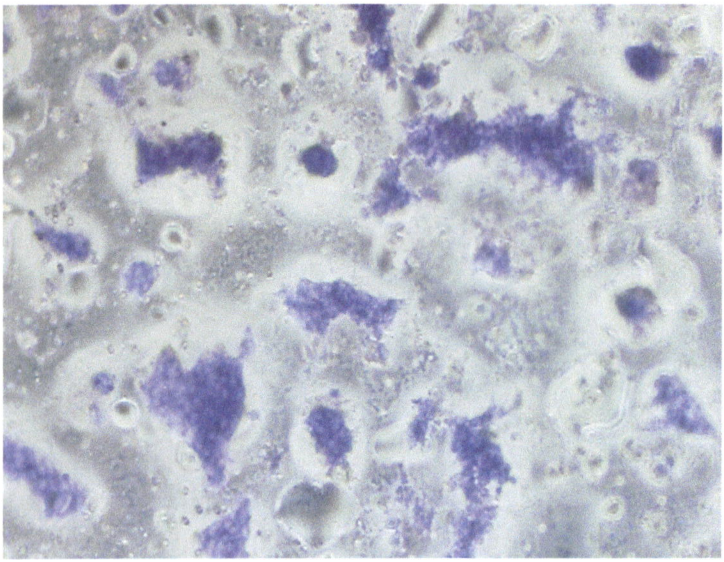

Fig. 14.7 Representative image of the invaded miPS-huh7cm cells in the invasion assay. Objective lens×20

26. Gently remove the remaining cells and the gel in the upper compartment of the insert with a cotton swab.
27. Put the insert in a new 24-well plate.
28. Fix the membrane with precooled 100% methanol for 10 min at room temperature.
29. Stain the invaded cells with Giemsa solution for an additional 30 min.
30. Wash the insert with PBS to remove excess dye.
31. Drain excess PBS using a cotton swab.
32. Dry the insert completely in the air.
33. Count the number of stained cells on the lower side of the filter under a microscope and take photographs of the invaded cells (Fig. 14.7).

Notes

1. FBS should be batch-tested to ensure optimal ES cell growth and viability before starting your experiments.
2. Pre-warm the culture medium at 37 °C before use.
3. Matrigel solution is usually liquid at 4 °C, but it becomes gel at RT. The mixture of Matrigel and DMEM should be fresh and prepared on ice just before use.
4. In case the pH is higher than expected, adjust the pH using KH_2PO_4, while lower than expected adjust the pH using Na_2HPO_4.
5. Divide the cells into approximate ratios of 1:2 and make different numbers of cells for seeding to avoid overgrowth.

6. Monitor the cells every day under an inverted microscope with bright images taking photos in 10×, 20×, or 40× so that the cell shapes could be confirmed during the passage.
7. This dish will be used at Step 7.
8. These cells will be used at Step 22.

References

Afify SM, Hassan G, Osman A, et al. Metastasis of cancer stem cells developed in the microenvironment of hepatocellular carcinoma. Bioengineering (Basel). 2019;6(3):73.

Afify SM, Hassan G, Seno A, Seno M. Cancer-inducing niche: the force of chronic inflammation. Br J Cancer. 2022;127(2):193–201.

Afify SM, Sanchez Calle A, Hassan G, Kumon K, Nawara HM, Zahra MH, Mansour HM, Khayrani AC, Alam MJ, Du J, Seno A, Iwasaki Y, Seno M. A novel model of liver cancer stem cells developed from induced pluripotent stem cells. Br J Cancer. 2020;122(9):1378–90.

Afify SM, Seno M. Conversion of stem cells to cancer stem cells: undercurrent of cancer initiation. Cancers (Basel). 2019;11(3):345.

Ajani JA, Song S, Hochster HS, Steinberg IB. Cancer stem cells: the promise and the potential. Semin Oncol. 2015;42:S3–S17.

Albini A. Tumor and endothelial cell invasion of basement membranes. The matrigel chemoinvasion assay as a tool for dissecting molecular mechanisms. Pathol Oncol Res. 1998;4(3):230–41.

Castellone RD, Leffler NR, Dong L, Yang LV. Inhibition of tumor cell migration and metastasis by the proton-sensing GPR4 receptor. Cancer Lett. 2011;312(2):197–208.

Dalerba P, Kalisky T, Sahoo D, Rajendran PS, Rothenberg ME, Leyrat AA, Sim S, Okamoto J, Johnston DM, Qian D, et al. Single-cell dissection of transcriptional heterogeneity in human colon tumors. Nat Biotechnol. 2011;29:1120–7.

Friedl P, Alexander S. Cancer invasion and the microenvironment: plasticity and reciprocity. Cell. 2011;147:992–1009.

Mansour H, Afify SM, Hassan G, et al. Comparative study of metastatic potentials of three different cancer stem cell models. Adv Cancer Biol Metastasis. 2022;5:100062.

Mansour H, Hassan G, Afify SM, Yan T, Seno A, Seno M. Metastasis model of cancer stem cell-derived Tumors. Methods Protoc. 2020;3(3):60.

Massagué J, Obenauf AC. Metastatic colonization by circulating tumour cells. Nature. 2016;529:298–306. https://doi.org/10.1038/nature17038.

Reymond N, d'Água BB, Ridley AJ. Crossing the endothelial barrier during metastasis. Nat Rev Cancer. 2013;13:858–70.

Chapter 15
Metastatic Potential of Cancer Stem Cells In Vivo

Abstract Metastasis is a major cause of cancer-related deaths more than 90% all over the world. Malignant tumor cells usually reach secondary sites via blood vessels. Metastatic process in solid tumors contains multiple steps that start from the invasion of cancer cell into basement membrane and followed by intravasation representing migration into the surrounding vasculature or lymphatic system. Then the cells survive in the blood flow and extravasation occurs from vasculature to secondary tissue. Finally, the metastasis is completed by colonization. During the last two decades scientists came to believe that cancer stem cells were responsible for the metastatic events taking epithelial to mesenchymal transition into consideration.

Nowadays, the concept of cancer stem cells is very critical to study the process of metastasis. Since cancer stem cells are available by the conversion of induced pluripotent stem cells in the presence of conditioned medium of cancer cells, the study on the mechanism of metastasis has become practically possible.

In this chapter, we evaluate the detailed steps of cancer stem cell preparation and injection into spleen and tail vein. This process will help understand the metastasis from the cancer stem cell perspective and give new ideas for the strategies of treatment in the future.

Keywords Hepatic Metastases · Lung metastasis CSCs · Intra-splenic transplantation

15.1 Introduction

Metastasis is considered as the development of secondary tumors in a site distant from the original site of the primary tumor. Metastasis is one of the most critical issues that lead to the failure of cancer treatment and the mortality of patients (Guan 2015, Fares et al. 2020). The steps of tumor progression, invasiveness, metastases, and recolonization should be understood in detail in order to develop more efficient therapies. To answer this purpose, successful establishment of metastasis models requires malignant cells, which have the potential to make invasion from their primary tumor into surrounding tissues followed by intravasation into blood vessels where the cells circulate in the blood flow, to get out of blood vessels by

S. M. Afify, M. Seno, *Methods in Cancer Stem Cell Biology*,
https://doi.org/10.1007/978-981-99-1331-2_15

extravasation into secondary site surrounded by different cellular environment, and to colonize forming secondary tumor (Maitra 2019; Massague and Obenauf 2016).

Several cancer models including metastasis have been proposed and established to develop the novel therapeutic agents. Animal models of metastasis have been established by either spontaneous or experimental procedure (Price 2014, Francia et al. 2011).

In the spontaneous model, cancer cells are injected into an animal at the specific sites forming a primary tumor followed by the spontaneous metastasis (Killion et al. 1998-1999). This model allows the observations of all the metastatic processes, including primary tumor formation, invasion, intravasation, extravasation, formation of micro metastasis, and colonization contributing to understand the metastasis, although their relevance to the biology of the multi-step metastatic process should be interpreted with caution.

In the experimental model, metastases are induced by direct intravasation of cancer cells injected into the vasculature skipping the steps of primary tumor growth. In this model, tumor cells are injected either intravenously or into highly vascularized organs such as the spleen. The site of injection may be chosen depending on the expected location of metastasis (Khanna and Hunter 2005). When injected into the lateral tail vein, the cells generally metastasize to the lung (Afify et al. 2020), while metastasizing to the liver by intrasplenic or portal vein injections (Afify et al. 2019).

In recent years, increasing evidence has demonstrated that CSCs are associated with aggressive characteristics, such as increased survival and resistance to treatment (Ajani et al. 2015, Afify and Seno 2019, Afify et al. 2022). In spite of the fact that the mechanism of metastasis is unclear, CSCs have been reported to contribute to the process (Afify et al. 2019, Mansour et al. 2020, Mansour et al. 2022). In the current chapter, we evaluate the detailed steps of cancer stem cell preparation and injection into the spleen and tail vein.

15.2 Materials

15.2.1 Reagents

- Cancer stem cells (e.g., miPS-Huh7cm cells) (Afify et al. 2019; Afify et al. 2020).
- Dulbecco's modified Eagle medium (DMEM, Sigma-Aldrich, St. Louis, MO, USA).
- Disodium hydrogen phosphate (Na2HPO4) (Sigma-Aldrich, cat. no. 255793).
- Fetal bovine serum (FBS, e.g., Gibco, Life Technologies, Massachusetts, USA) (Note 1).
- Penicillin/streptomycin mixed solution (Nacalai Tesque, Japan).
- Phosphate-buffered saline (Dulbecco's formula PBS) (Genesee Scientific, El Cajon, USA).
- Potassium chloride (KCl) (Sigma-Aldrich, cat. no. P9333).

- Potassium dihydrogen phosphate (KH2PO4) (Sigma-Aldrich, cat. no. P0662).
- MEM non-essential amino acids solution (×100) (Thermo Fisher Scientific, Waltham, MA, USA).
- L-Glutamine (Nacalai Tesque, Japan).
- Sodium chloride (NaCl) (Sigma-Aldrich, cat. no. 746398).
- 2-Mercaptoethanol (Sigma-Aldrich, St. Louis, USA).
- 2.25% Trypsin-EDTA (Atlanta Biologicals, Flowery Branch, USA).
- Gelatin, tissue culture grade (type B, 2% in H_2O; Sigma-Aldrich, cat. no. G1393).

15.2.2 Equipment

- Adjustable pipettes: P-20 (Gilson, cat. no. FA10003M), P-200 (Gilson, cat. no. FA10005M) and P-1000 (Gilson, cat. no. F123602M).
- Biosafety Cabinet (BSC-04IIA2) airtech.
- Safety Cabinet Class 2 (ESCO, model no. LA2-4A1).
- Cell culture incubator (CO2 at 5%, humidified at 37 °C) (Thermo Scientific cat. no. 311).
- CKX41 Inverted Microscope – Olympus Life Science (Japan).
- Cotton swabs (Sigma-Aldrich, cat. no. Z699365).
- Eppendorf 5415R Refrigerated Centrifuge (Eppendorf, Germany).
- Falcon® Conical Centrifuge Tubes (15 mL; BD Falcon, New York, NY, USA, Cat. No 352095).
- Hemocytometer (Sigma-Aldrich, cat. no. Z359629).
- Laser scanning confocal microscope, FV-1000, (Olympus, Tokyo, Japan).
- Liquid N2 storage tank.
- Microcentrifuge 1.5-mL tubes (Eppendorf, cat. no. 0030120086).
- Pipette tips 20 µl (Starlab, cat. no. S1120–1810), Pipette tips: 200 µl (Starlab, cat. no. S1120–8810), Pipette tips 1000 µl (Starlab, cat. no. S1120–1830).
- Sterile forceps.
- Syringes.
- Software/ ImageJ – Computer Vision Online.
- Water bath (Thermo Scientific cat. no. 152–4101).
- Tissue culture-treated plate, 60 mm dish (TPP Techno Plastic Products AG Schaffhausen, Switzerland, Cat. No 93060).
- Tissue culture-treated plate, 100 mm dish (TPP Techno Plastic Products AG Schaffhausen, Switzerland, Cat. No 93100).
- Transwell inserts for 24-well plates (membrane 8.0 µm pores) (Corning, cat. no. 353097).
- 5/10/25-mL plastic disposable pipette.

15.2.3 Reagent Preparation

15.2.3.1 Medium for Cancer Stem Cells Culture

For cancer stem cells, prepare high glucose containing DMEM supplemented with final concentrations of 15% FBS, 0.1 mM MEM non-essential amino acids solution, 2 mM L-Glutamine, 0.1 mM 2-mercaptoethanol, 50 U/mL penicillin/streptomycin. Store at 4 °C (Note 2).

15.2.3.2 PBS (pH 7.40) Preparation

Prepare PBS by mixing 137 mM NaCl, 10 mM Na2HPO4, 2.7 mM KCl, and 1.8 mM KH2PO4. Prepare the buffer in distilled water (dH2O), adjust the pH to 7.4 and autoclave it for 20 min (Note 4).

15.2.3.3 0.1%(W/V) Gelatin

Dissolve 0.5 g of gelatin (from porcine skin) in 500 ml distilled water and autoclave. Store at 4 °C.

15.3 Methods

This chapter describes the intrasplenic transplantation of CSCs as the model of hepatic metastasis step by step as well as lung metastasis by injection in vein. In the current evaluation miPS-Huh7cm cells were used as CSCs which were developed in the microenvironment of hepatocellular carcinoma (Fig. 15.1).

This protocol will be very important to guide researchers who will investigate molecular mechanisms of metastasis in liver developing from CSCs and screen new therapeutic drugs. The surgery should be performed under sterile conditions. All instruments should be autoclaved in advance.

15.3.1 Preparation of Gelatin-Coated Dishes

1. Transfer 0.2% gelatin solution from 4 °C to the clean bench.
2. Cover the bottom of the dish with sufficient volume of 0.2% gelatin solution.
3. Incubate the dishes for 30 min at 37 °C in an incubator with 5% CO_2.

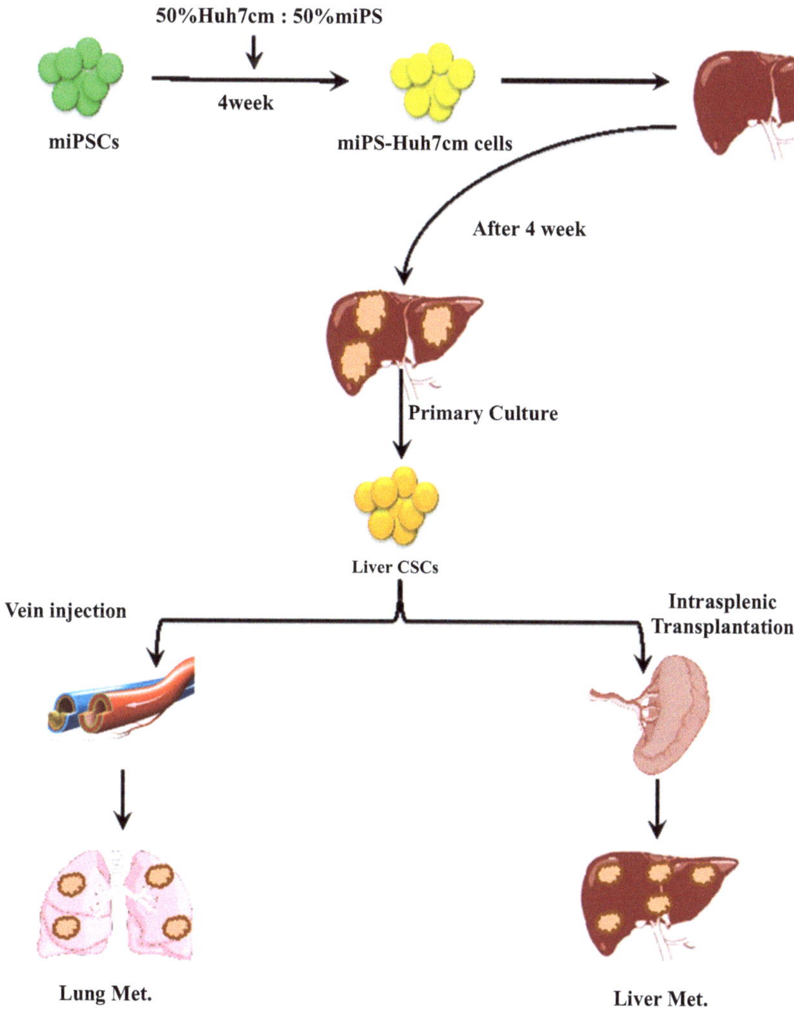

Fig. 15.1 Representative scheme for liver CSCs generation and spleen/vein injection

15.3.2 Preparation of Cancer Stem Cells

Recently, Afify et al., generated CSCs from iPSCs/ESCs (Afify et al. 2019). iPSCs were exposed to the conditioned medium (CM) of human hepatocellular carcinoma cell line Huh7 cells for 4 weeks to induce liver CSCs (miPS-Huh7cm cells) (Afify et al. 2019; Afify et al. 2020). Reviving and passaging CSCs should be referred to the previous chapter in detail. Briefly, overview the following steps.

1. Culture a sufficient number of CSCs in high-glucose DMEM containing 15% FBS in 60-mm tissue culture dishes in a humidified 37 °C, 5% CO_2 incubator.

2. Aspirate the medium from sub-confluent cultures.
3. Add 2 ml 0.025% trypsin-EDTA solution per dish.
4. Incubate the dish for 2–3 min or less at 37 °C until the cell shape becomes round (Note 1).
5. Pour 5 ml of warm medium in the dish.
6. Transfer the cell suspension into a sterile 15-ml falcon tube.
7. Centrifuge 5 min at ~200 × g at room temperature.
8. Discard the supernatant and suspend the cell ppt in 1 mL of fresh warmed medium.
9. Count cells on a hemocytometer.
10. Prepare $1x10^6$ cell/1.5-mL tube suspended in 100 μL of HBSS (Note 2).

15.3.3 Intra-Splenic Transplantation of Cancer Stem Cells

Intrasplenic cell injection is currently considered as the "gold standard" procedure for cell transplantation in order to develop liver metastasis. With a scalpel blade, skin was first incised at the top left of the back of a mouse over the spleen, the subcutaneous layer was cut, and muscle layer was opened so as to find the spleen. The spleen was mobilized and translocated onto the subcutaneous layer. Cancer stem cells were very slowly injected into the spleen. The spleen was then relocated back to the normal position. Finally, the skin was stitched by a surgical suture with a needle. Liver metastasis was developed within 1 month after injection (Fig. 15.2).

1. Autoclave all surgical materials, suture thread, suture needle, and cotton.
2. Set all surgical materials in the laminar hood.
3. Turn on the UV lamp for 20 min on all materials.
4. Turn off the UV lamp and turn on the light.
5. Anesthetize the animals with 2% isoflurane (Note 3).
6. Check if the mouse is completely anesthetized by pinching the toe without withdrawal reflex.
7. Place the left side back of the mice upwards.
8. Disinfect the left side area over the spleen with 70% ethanol followed by iodine.
9. Make an incision with 2–3 cm length using a scalpel blade so that you can see the spleen (Fig. 15.2).
10. Lift the spleen carefully and locate it on the subcutaneous layer with forceps.
11. Draw up 50 μl of 5.0×10^5 miPS-Huh7cm cells in HBSS (from the last section) into a 26 G × 5/8″ syringe.
12. Insert the syringe needle into the tip of the spleen.
13. Inject the cells slowly into the spleen until it is completely injected (Note 4).
14. Wait for 60 sec (Note 5).
15. Remove the needle slowly.
16. Relocate the spleen into the cavity very carefully.
17. Stitch the incision by a surgical suture with needle (Note 6).
18. Disinfect the stitch of surgery by iodine and put the mice back in the cage.

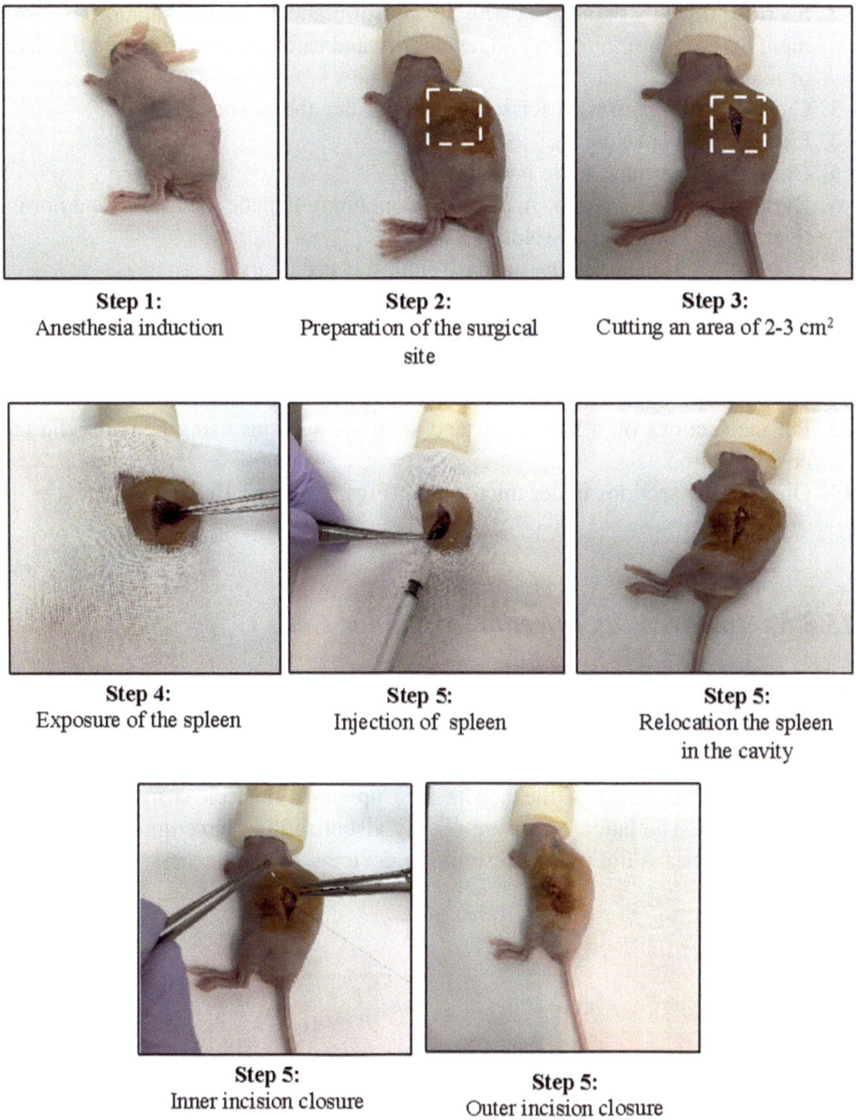

Fig. 15.2 Representative scheme for steps of intrasplenic transplantation of CSCs derived from miPSCs in the presence of hepatocellular carcinoma

15.3.4 Detection of Hepatic Metastasis

1. Between 20 and 30 days from transplantation, mice will begin to show clinical symptoms of metastasis and will require euthanasia (Note 7).

2. Sacrifice mice by euthanasia with 5% of isoflurane through inhalation to ensure rapid loss of consciousness and respiratory and cardiac arrest to ensure the death of mice.
3. Cut the peritoneum with scissors to expose the abdomen.
4. Excise the liver (Note 8).
5. Check the liver metastatic nodules.
6. Divide the liver lobe into small pieces containing hepatic metastasis and normal liver in the same section (Note 9).
7. Take some aliquots for the primary culture if necessary.
8. Fix the tissues in 10% formalin solution for paraffin embedded sections.
9. Allow the fixation for 24–48 h at 4 °C.
10. Embed the fixed tissue in paraffin.
11. Cut the embedded tissue into sections with 5 μm thickness.
12. Put the sections on a slide glass and stain the sections using hematoxylin and eosin solution.
13. Observe the sections under microscope (Figs. 15.3 and 15.4).

15.3.5 *Intravenous Injection of Tumor Cells*

The mouse is either placed in the restrainer or anesthetized and the tail is immersed in warm water (40–45 °C) in order to dilate the vessels. The tail is swabbed with 70% alcohol. The needle is inserted parallel to the tail vein penetrating into the lumen while keeping the bevel of the needle face upward. The cell suspension is then injected slowly. The lateral veins are readily visualized but have quite small diameters. If anesthesia is not used, a restraining device is usually necessary.

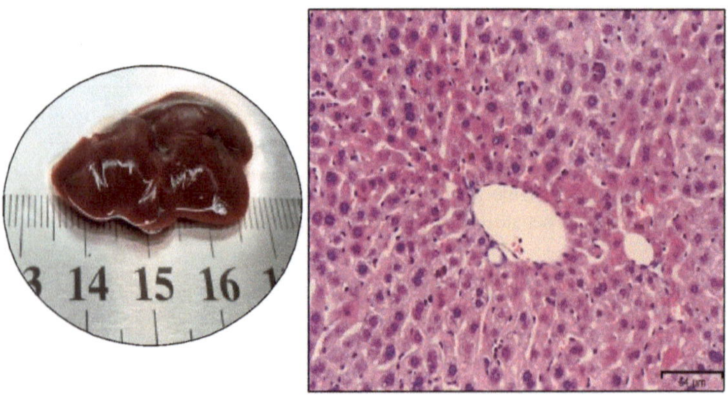

Fig. 15.3 Examination of the liver following spleen transplantation with HBSS without any cells as control

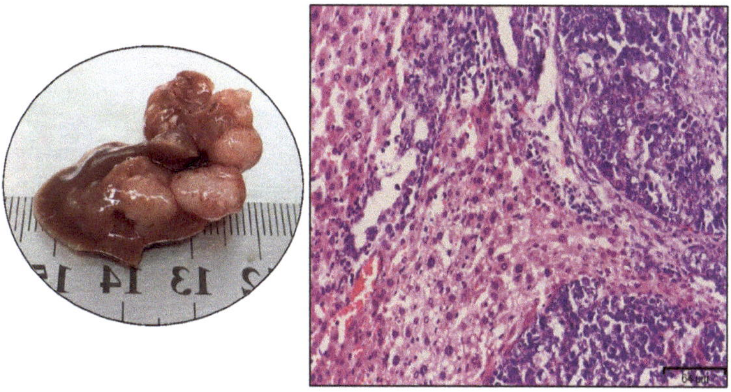

Fig. 15.4 Examination of the liver following spleen transplantation with CSCs showed liver metastases that replace the normal hepatic parenchyma compared to normal adult liver. Scale bar 64 μm

1. Take the mouse out of the cage grasping the tail.
2. Dilate the lateral tail veins by immersing the tail for 1–2 min in hot water or fix the tail on a hot plate at approximately 50–60 °C (Note 10).
3. Place the mouse in a restrainer having the tail protruded through the hole in the wall of the device.
4. Draw up 100 μl of 1.0×10^6 miPS-Huh7cm cells in HBSS (from the last section) into a 26 G × 5/8″ syringe.
5. Remove air bubbles from the syringe (Note 11).
6. Hold the distal third part of the tail with fingers.
7. Disinfect the tail with 70% ethanol and rotate the position to the lateral vein.
8. Stabilize the vein with fingers.
9. Apply slight pressure to straighten the tail and further dilate the lateral vein.
10. Hold the 1-mL syringe with 26 G x 5/8″ needle filled with the cells in the other hand.
11. Push the needle gently through the skin (at a slight angle) into the vein.
12. Advance the needle into the lumen for an additional 5 mm upon cannulation and slowly inject 0.1 mL of the cell suspension (Note 12).
13. Remove the needle and hold the hole of the injection tail with the fingers for 30 sec to avoid backflow.
14. Put the mouse back in the cage.
15. Direct the bevel of the needle upward and discard.

15.3.6 Detection of Lung Metastasis

Sacrifice mice by euthanasia with 5% of isoflurane through inhalation to ensure rapid loss of consciousness and respiratory and cardiac arrest to ensure the death of mice. Note: The mouse will require euthanasia between 3 and 4 weeks from injection depending on the condition of the mouse and the phenotype of the injected cells.

1. Cut the position along with the diaphragm with scissors to expose the lungs.
2. Excise the lung with the heart (Note 13).
3. Check the lung metastatic nodules.
4. Divide the lung lobe into small pieces containing lung metastasis (Note 14).
5. Take some aliquots for the primary culture if necessary.
6. Fix the tissues in 10% formalin solution for paraffin-embedded sections.
7. Allow the fixation for 24–48 h at 4 °C.
8. Embed the fixed tissue in paraffin.
9. Cut the embedded tissue into sections with 5 μm thickness.
10. Put the sections on a slide glass and stain the sections using hematoxylin and eosin solution.
11. Observe the sections under microscope (Fig. 15.5).

Notes

1. No need to wait for the cells detached from the bottom.
2. Half of this volume will be used at Step 11 in the next Sect. 15.2.1 and Step 12 in the section after the next. Be sure to prepare more than the volume you need so that Pipetting errors should be avoided.
3. Take care not to breathe isoflurane with some suitable gas trapping apparatus.
4. Careful attention to be paid to the slowness in the infusion of cells.
5. Do not remove the needle immediately after infusion to avoid back flow of cells as well as bleeding. After 3 min, clotting will appear at the site of puncture.
6. Irrigate the incision with distilled water so that the skin is kept wet.
7. Swelling in the abdominal cavity and significant weight loss will be the signs.
8. Put the liver in a 100-mm dish containing HBSS.
9. Take small pieces for RNA in Liq -N2.
10. The tail vein will be dilated to facilitate the insertion of a 26- or 27-G needle.

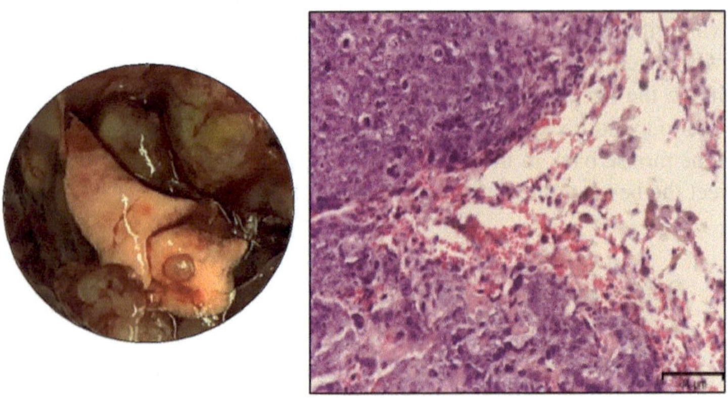

Fig. 15.5 Examination of the lung following tail vein injection with CSCs. Scale bar 64 μm

11. Avoid the injection of air into the vein.
12. The cells are from Step 10 in the previous section of cell preparation.
13. Put the tissues in a 100-mm dish containing HBSS.
14. Take small pieces for RNA in Liq -N2.

Acknowledgments We would like to show our gratitude to the Japan Society for the Promotion of Science (JSPS) for supporting this research at Okayama University.

References

Afify SM, Hassan G, Osman A, et al. Metastasis of cancer stem cells developed in the microenvironment of hepatocellular carcinoma. Bioengineering (Basel). 2019;6(3):73.

Afify SM, Hassan G, Seno A, Seno M. Cancer-inducing niche: the force of chronic inflammation. Br J Cancer. 2022;127(2):193–201.

Afify SM, Sanchez Calle A, Hassan G, Kumon K, Nawara HM, Zahra MH, Mansour HM, Khayrani AC, Alam MJ, Du J, Seno A, Iwasaki Y, Seno M. A novel model of liver cancer stem cells developed from induced pluripotent stem cells. Br J Cancer. 2020;122(9):1378–90.

Afify SM, Seno M. Conversion of stem cells to cancer stem cells: undercurrent of cancer initiation. Cancers (Basel). 2019;11(3):345.

Ajani JA, Song S, Hochster HS, Steinberg IB. Cancer stem cells: the promise and the potential. Semin Oncol. 2015;42:S3–S17.

Fares J, Fares MY, Khachfe HH, et al. Molecular principles of metastasis: a hallmark of cancer revisited. Sig Transduct Target Ther. 2020;5:28. https://doi.org/10.1038/s41392-020-0134-x.

Francia G, Cruz-Munoz W, Man S, Xu P, Kerbel RS. Mouse models of advanced spontaneous metastasis for experimental therapeutics. Nat Rev Cancer. 2011;11(2):135–41. https://doi.org/10.1038/nrc3001.

Guan X. Cancer metastases: challenges and opportunities. Acta Pharm Sin B. 2015;5(5):402–18.

Khanna C, Hunter K. Modeling metastasis in vivo. Carcinogenesis. 2005;26(3):513–23.

Killion JJ, Radinsky R, Fidler IJ. Orthotopic models are necessary to predict therapy of transplantable tumors in mice. Cancer Metastasis Rev. 1998-1999;17(3):279–84.

Maitra A. Molecular envoys pave the way for pancreatic cancer to invade the liver. Nature. 2019;567:181–2.

Mansour H, Afify SM, Hassan G, et al. Comparative study of metastatic potentials of three different cancer stem cell models. Advances in Cancer Biology – Metastasis. 2022;5:100062.

Mansour H, Hassan G, Afify SM, Yan T, Seno A, Seno M. Metastasis model of cancer stem cell-derived Tumors. Methods Protoc. 2020;3(3):60.

Massague J, Obenauf AC. Metastatic colonization by circulating tumour cells. Nature. 2016;529:298–306.

Price JE. Spontaneous and experimental metastasis models: nude mice. Methods Mol Biol. 2014;1070:223–33.

Chapter 16
Anchorage-Independent Cell Growth Assay for Cancer Stem Cells: Tumorigenic Assay in Vitro

Abstract Anchorage-independent growth is a characteristic feature of transformed cells, which means the cells have the ability to grow independently of a solid surface. This potential of the cell growth in vitro is generally referred to the tumorigenicity as a property of cancer cells. The soft agar colony formation assay is a well-established method to evaluate this capability in vitro and is considered as one of the most stringent tests to distinguish cells transformed into malignant cells from normal cells. This evaluation of phenotype is assumed to be advantageous to that in vivo because in vitro assays should be easier and economical without sacrificing animals. Furthermore, sometimes malignant cells are not tumorigenic due to the difference of species between the cells and the host as in the cases of human cancer cells transplanted in mice due to the host defense by the immunological system.

In the current chapter, we will assess the soft agar colony formation assay using CSC models derived from miPSCs. Since CSCs are distinguished from the ordinary cancer cells, in vitro growth independent of a solid surface could be an effective way to evaluate different therapeutic agents targeting CSC in the future.

Keywords Cancer stem cells · Soft agar colony · Cancer model

16.1 Introduction

Neoplastic cell is considered to be generated by the transformation due to the accumulation of genetic changes resulting in uncontrolled growth without following homeostatic regulation. The alterations will be obvious in cell surface, morphology and metabiological diversity, karyotype with abnormalities, and other attributes conferring the ability to invade and metastasize (Afify et al. 2022). Carcinogenesis is a complicated process composed of numerous steps and factors (Afify and Seno 2019). Many genes and signaling pathways have experimentally been identified and characterized as the oncogenic phenotype (Afify et al. 2021; Chen et al. 2022). Actually, the two different studies in the genes and in the signaling pathways appeared to facilitate a better understanding of cancer.

The anchorage-independent colony formation is one of the hallmarks that was found through the experiments recognized as the characteristics of malignancy

S. M. Afify, M. Seno, *Methods in Cancer Stem Cell Biology*,
https://doi.org/10.1007/978-981-99-1331-2_16

in vitro residing in the cells transformed with uncontrolled cell growth (Borowicz et al. 2014). Normal epithelial cells generally survive and proliferate dependent on the basement membranes in adhesive conditions while they cause apoptosis when transferred into suspension culture. In contrast, malignant cells will form colonies in an attachment-independent manner keeping proliferation without apoptosis.

Traditionally, soft agar colony formation assay is used to assess the anchorage-dependent cell growth (Puck et al. 1956). The cells are grown in a layer of soft agar mixed with cell culture medium that rests on another layer of agar, which prevents cells from adhering to the culture plate and allows transformed cells to form visible colonies.

Thus, by the soft agar colony formation assay you can measure the proliferation of cells by manually counting the colonies formed on a semisolid condition after 3–4 weeks.

This stringency makes this assay as a superior tool to evaluate the effectiveness of drug candidates (Roberts et al. 1985, Freedman and Shin 1974, Horibata et al. 2015, Anderson et al. 2007). Furthermore, in this method cells grow as colonies recapitulating the physiological environment in vivo in terms of cellular interactions and unique growth conditions.

Hence, the soft agar assay offers key advantages more than conventional 2D assays used for drug screening (Weaver et al. 2002, Bhadriraju and Chen 2002, Fukazawa et al. 2002). However, standard soft agar assay is a kind of low-throughput with manual counting of cell colonies and colony volumes. Although this sounds far from high-throughput drug screens, the soft agar assay is useful generally to assess the tumorigenicity in vitro.

This method could be used to distinguish each different subpopulation in a tumor tissue because limited subpopulations are considered to have the ability to proliferate forming spheres. The sphere-forming subpopulation is supposed to consist of CSC while the rest can only show low potential. The evaluation of this method with CSCs developed from iPSCs (Chen et al. 2012) could be very important because it could be used to evaluate the effect of different therapeutic agents against CSCs (Xu et al. 2022).

16.2 Materials

16.2.1 Reagents

- Cancer stem cells (e.g., miPS-LLCcm cells) (Chen et al. 2012).
- Dulbecco's modified Eagle medium (DMEM, Sigma-Aldrich, St. Louis, MO, USA).
- Disodium hydrogen phosphate (Na2HPO4) (Sigma-Aldrich, cat. no. 255793).
- Fetal bovine serum (FBS, e.g., Gibco, Life Technologies, Massachusetts, USA).
- Penicillin/streptomycin mixed solution (Nacalai Tesque, Japan).

- Phosphate-buffered saline (Dulbecco's formula PBS) (Genesee Scientific, El Cajon, USA).
- Potassium chloride (KCl) (Sigma-Aldrich, cat. no. P9333).
- Potassium dihydrogen phosphate (KH2PO4) (Sigma-Aldrich, cat. no. P0662).
- MEM non-essential amino acids solution (×100) (Thermo Fisher Scientific, Waltham, MA, USA).
- L-Glutamine (Nacalai Tesque, Japan).
- Sodium chloride (NaCl) (Sigma-Aldrich, cat. no. 746398).
- 2-Mercaptoethanol (Sigma-Aldrich, St. Louis, USA).
- 2.25% Trypsin-EDTA (Atlanta Biologicals, Flowery Branch, USA).
- Gelatin, tissue culture grade (type B, 2% in H2O; Sigma-Aldrich, cat. no. G1393).
- Isoflurane (gas anesthesia system) DS Pharma Animal Health Co., Ltd. Osaka 541-0053, Japan).
- BALB/c-nu/nu, female, 4 weeks old, Charles River laboratories, Kanagawa, Japan.

16.2.2 Equipment

- Adjustable pipettes: P-20 (Gilson, cat. no. FA10003M), P-200 (Gilson, cat. no. FA10005M), and P-1000 (Gilson, cat. no. F123602M).
- Biosafety Cabinet (BSC-04IIA2) airtech.
- Safety Cabinet Class 2 (ESCO, model no. LA2-4A1).
- Cell culture incubator (CO2 at 5%, humidified at 37 °C) (Thermo Scientific cat. no. 311).
- CKX41 Inverted Microscope – Olympus Life Science (Japan).
- Cotton swabs (Sigma-Aldrich, cat. no. Z699365).
- Eppendorf 5415R Refrigerated Centrifuge (Eppendorf, Germany).
- Falcon® Conical Centrifuge Tubes (15 ml; BD Falcon, New York, NY, USA, Cat. No 352095).
- Hemocytometer (Sigma-Aldrich, cat. no. Z359629).
- Laser scanning confocal microscope, FV-1000, (Olympus, Tokyo, Japan).
- Liquid N2 storage tank.
- Microcentrifuge 1.5-ml tubes (Eppendorf, cat. no. 0030120086).
- Pipette tips 20 μl (Starlab, cat. no. S1120–1810), Pipette tips: 200 μl (Starlab, cat. no. S1120–8810), Pipette tips 1000 μl (Starlab, cat. no. S1120-1830).
- Sterile forceps.
- Syringes.
- Software/ ImageJ – Computer Vision Online.
- Water bath (Thermo Scientific cat. no. 152–4101).
- Tissue culture–treated plate, 60 mm dish (TPP Techno Plastic Products AG Schaffhausen, Switzerland, Cat. No 93060).

- Tissue culture–treated plate, 100 mm dish (TPP Techno Plastic Products AG Schaffhausen, Switzerland, Cat. No 93100).
- Transwell inserts for 24-well plates (membrane 8.0 μm pores) (Corning, cat. no. 353097).
- 5/10/25-ml plastic disposable pipette.
- Microwave or heating block.

16.2.3 Reagent Preparation

Stem Cell Medium

Mix the following for 500 ml		
Dulbecco's modified Eagle's medium-high glucose		412.5 ml
FBS		75 ml
Penicillin/streptomycin mixed solution	(50 U/ml)	2.5 ml
Gibco™ L-Glutamine (200 mM)	(2 mM)	5 ml
MEM non-essential amino acids solution	(1 mM)	5 ml

PBS (pH 7.40) Preparation

Prepare PBS by mixing 137 mM NaCl, 10 mM Na2HPO4, 2.7 mM KCl, and 1.8 mM KH2PO4. Prepare the buffer in distilled water (dH2O), adjust the pH to 7.4, and autoclave it for 20 min.

0.1%(w/v) Gelatin

Dissolve 0.5 g of gelatin (from porcine skin) in 500 ml distilled water and autoclave. Store at 4 °C.

1.2% Agar Solution

Place 1 g of agar powder in a sterile bottle, add 100 ml of sterile cell culture grade water. Microwave or boil until agar is completely dissolved.

0.6% Agar Solution

Place 0.3 g of agar powder in a sterile bottle, add 100 ml of sterile cell culture grade water. Microwave or boil until agar is completely dissolved. When stored at 2–8 °C, the agar and cell stain powder are stable up to the expiration date. Cell stain quantification solution should be stored at −20 °C up to expiration date. Discard any remaining reagents after the expiration date.

16.3 Methods

The soft agar assay for colony formation is an evaluation of anchorage-independent growth of cells in soft agar. In this assay, cells are cultured in soft agar medium for 14 days. Then the tumorigenicity of the cells can be analyzed by the morphology and quantified by the number of colonies formed. All the following steps must be performed in a standard tissue culture hood in an aseptic way. Contamination and overgrowth should be avoided all through the experiment. Cells should be passaged at least 3 times before starting the experiment as following steps in Fig. 16.1.

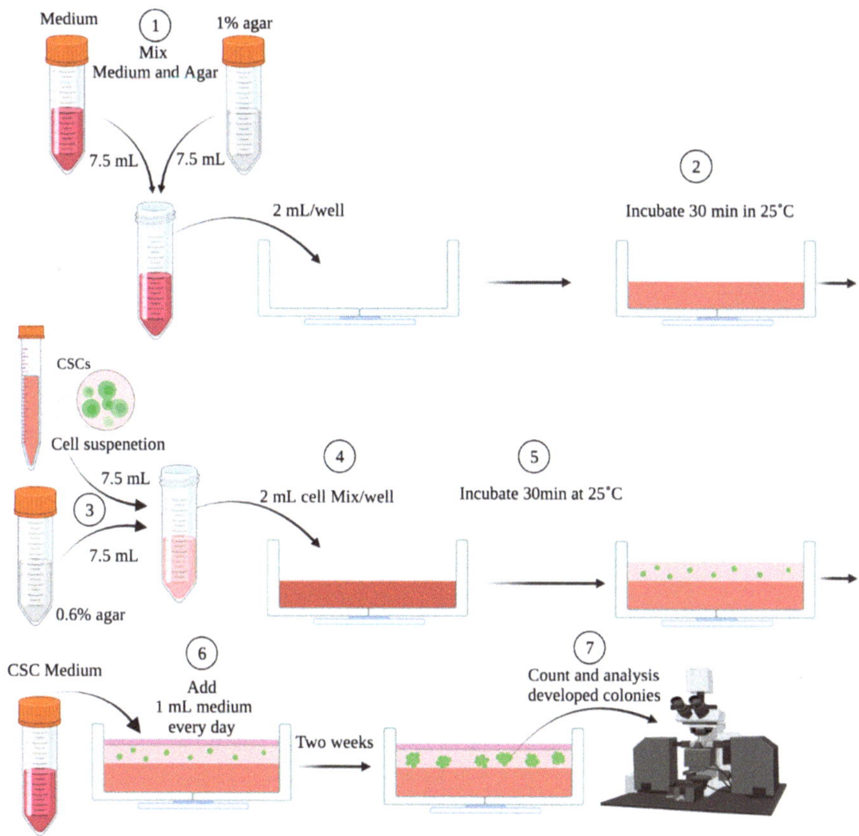

Fig. 16.1 Representative scheme for anchorage-independent cell growth assay with CSCs induced from iPSCs

16.3.1 Preparation of Gelatin-Coated Dishes

1. Transfer 0.2% gelatin solution from 4 °C to the clean bench.
2. Cover the bottom of the dish with sufficient volume of 0.2% gelatin solution.
3. Incubate the dishes for 30 min at 37 °C in an incubator with 5% CO_2.

16.3.2 Cancer Stem Cell Thawing

1. Coat three 60-mm dishes with 0.1% gelatin at 37 °C in an incubator balanced with 5% CO_2 for at least 30 minutes.
2. Pre-warm the medium for cancer stem cell culture at 37 °C.
3. Transfer 5 ml of pre-warmed medium to a sterile 15-ml conical tube.
4. Pick up a vial of miPS-LLCcm cells from liq-N_2.
5. Thaw it in a 37 °C water bath incubating for 1–3 min.
6. Sterilize the vial with cotton in 70% ethanol.
7. Put the vial of cells inside of the cell culture hood.
8. Transfer the cells into a sterilized 15-ml conical tube.
9. Add 5 ml of pre-warmed medium prepared at Step 3 slowly into the tube dropping one by one with gentle mixing.
10. Centrifuge the tube at 200 × g for 5 min.
11. Remove the supernatant without disturbing the cell pellet.
12. Resuspend the cell pellet in 1 ml of pre-warmed medium.
13. Seed the cells on three gelatin-coated 60-mm dishes.
14. Incubate the cells at 37 °C in an incubator balanced with 5% CO_2.
15. Change medium at 2-day intervals observing the cell conditions every day. Note 1.
16. Observe the cells with a GFP microscope.
17. Treat cells overnight with puromycin (1 μg/ml) (Fig. 16.2).

16.3.3 Preparation of Base Agar Layer

1. Pre-warm the culture medium in a bucket filled with warm tap water at 42 °C.
2. Loosen the cap on the bottle of 1% agar.
3. Melt the agar in a microwave for at least 3 min. Note 1.
4. Cool the melted agar solution to 37–42 °C in a water bath. Note 2.
5. Place a sterile 50-ml empty conical tube in a tube holder in the bucket with warm water.
6. Transfer bucket to cell culture clean bench.
7. Transfer 7.5 ml of culture medium to the 50 ml conical tube and then 7.5 ml of the 1% noble agar solution on the warm water. Note 3.

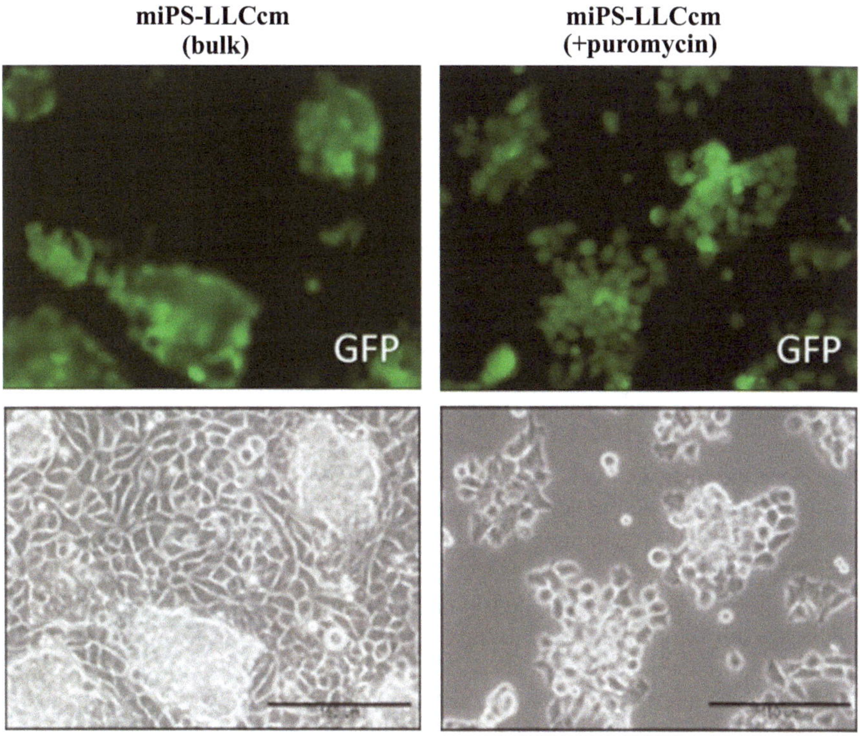

Fig. 16.2 Microscopic images of miPS-LLcm cells maintained in complete stem cell medium (SCM) with and without puromycin (1 μg/ml)

8. Invert the conical tube many times to homogenize the mixture.
9. Transfer 2 ml of mixture into each well of a 6-well plate. Note 4.
10. Cover the 6-well plate and let stand at room temperature in the clean bench for at least 30 min allowing the base agar layer to solidify.

16.3.4 Cancer Stem Cell Preparation

1. From 70% confluent CSCs (miPS-LLCm cells), aspirate the medium in the culture dish.
2. Rinse the cells in the dish twice with sterile PBS.
3. Add 2 ml of 0.05% trypsin to the dish.
4. Incubate the dish for 5 min at 37 °C.
5. Stop trypsinization by adding FBS containing a pre-warmed medium.
6. Harvest the cells and transfer to a 15-ml conical tube.
7. Spin down the cells at 200 g for 5 min at 4 °C.
8. Aspirate the supernatant without disturbing the cell pellet.

9. Resuspend the pellet in 1 ml medium.
10. Determine the cell count with a hemocytometer.
11. Seed the cells at 0.3×10^6 on a gelatin-coated dish.
12. Resuspend the cells in a 15 ml tube.
13. Adjust the cell count in the suspension for seeding 5000 cells/per well.

16.3.5 Plating the Upper Layer of Agar Containing Cells

1. Pre-warm the culture medium in a bucket filled with tap water at 42 °C.
2. Melt 0.6% agar solution in a microwave. Note 5.
3. Cool the melted agar solution down to 37 °C in a water bath.
4. Place an empty 50-ml conical tube in a tube-holder in the bucket with water at 42 °C.
5. Place the tube with cell suspension from the previous section and the tube of 0.6% agar in the tube-holder in the bucket with water at 42 °C.
6. Place the bucket containing water at 42 °C along with the tubes in a tube-holder in the clean bench.
7. Mix 0.6% agar and cell suspension in a 1:1 ratio. Note 6.
8. Keep the temperature of this mixture at around 37 °C to avoid premature hardening with the cell that survived.
9. Transfer 2 ml of the mix of 0.6% agar and medium per well of a 6-well plate for the upper layer of agar. Note 7.
10. Keep the plate at 25 °C for 30 min in a clean bench allowing the cell/agar mixture to solidify.
11. Add 500 μL of culture medium to each well. Note: This layer should be maintained over the upper layer of agar to prevent it from drying.
12. Incubate the plate at 37 °C under 5% CO_2 in a humidified cell culture incubator for 2 weeks.
13. Add 500 μL of the medium twice a week.
14. Observe the colony formation under a light microscope (Fig. 16.3). Using this method, we could evaluate the effect of IPI549 and Gefitinib on cancer stem cells (Xu et al. 2022).

Notes

1. Monitor the solution until agar is completely dissolved and the solution is clear and avoid boiling over. Alternatively, you can use 1% agar directly after autoclaving.
2. Work quickly before agar starts gelation.
3. The ratio of agar and medium should be 1:1. Prepare 15 ml in total for all wells of 6-well plate because you need 2 ml per well.
4. These steps from 7 to 9 should be done quickly without trapping any air bubbles in the wells.

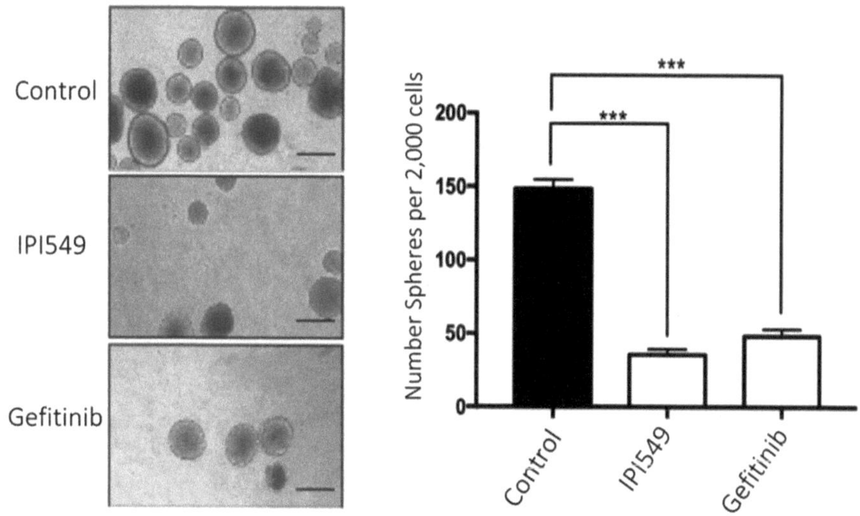

Fig. 16.3 MiPS-CSC models treated with Eganelisib and Gefitinib for 48 hours yielded spheres with a diameter greater than 75 mm. Based on 2000 seeded cells, the number of defined mammospheres was calculated. Each group was represented by bright field images. Scale bar: 100 μm. The figure is copied from our original paper: Xu, Y., Afify, S.M., Du, J. et al. The efficacy of PI3Kγ and EGFR inhibitors on the suppression of the characteristics of cancer stem cells. Sci Rep 12, 347 (2022)

5. Monitor the solution until agar is completely dissolved confirming the solution becomes clear but avoid boiling over.
6. First pipette 7.5 ml of cell suspension into the 50-ml conical tube then add 7.5 ml of 0.6% agar to the tube. Prepare a total mix of 15 ml in a 50-ml tube.
7. Pay attention not to trap any air bubbles into the wells.

References

Afify SM, Hassan G, Seno A, Seno M. Cancer-inducing niche: the force of chronic inflammation. Br J Cancer. 2022;127(2):193–201.

Afify SM, Oo AKK, Hassan G, Seno A, Seno M. How can we turn the PI3K/AKT/mTOR pathway down? Insights into inhibition and treatment of cancer. Expert Rev Anticancer Ther. 2021;21(6):605–19.

Afify SM, Seno M. Conversion of stem cells to cancer stem cells: undercurrent of cancer initiation. Cancers (Basel). 2019;11(3):345.

Anderson SN, Towne DL, Burns DJ, Warrior U. A high throughput soft agar assay for identification of anticancer compound. J Biomol Screen. 2007;12:938–45.

Bhadriraju K, Chen CS. Engineering cellular microenvironments to improve cell-based drug testing. Drug Discov Today. 2002;7:612–20.

Borowicz S, Van Scoyk M, Avasarala S, Rathinam MKK, Tauler J, Bikkavilli RK, Winn RA. The soft agar colony formation assay. J Vis Exp. 2014;92:e51998.

Chen L, Kasai T, Li Y, Sugii Y, Jin G, Okada M, Vaidyanath A, Mizutani A, Satoh A, Kudoh T, Hendrix MJ, Salomon DS, Fu L, Seno M. A model of cancer stem cells derived from mouse induced pluripotent stem cells. PLoS One. 2012;7(4):e33544.

Chen L, Liu Y, Xu Y, Afify SM, Gao A, Du J, Liu B, Fu X, Liu Y, Yan T, Zhu Z, Seno M. Up-regulation of Dsg2 conferred stem cells with malignancy through wnt/β-catenin signaling pathway. Exp Cell Res. 2022;113416

Freedman VH, Shin S-I. Cellular tumorigenicity in nude mice: correlation with cell growth in semi-solid medium. Cell. 1974;3:355–9.

Fukazawa H, Noguchi K, Murakami Y, Uehara Y. Mitogen-activated protein/ExtracellularSignal-regulated kinase kinase (MEK) inhibitors restore Anoikis sensitivity in human breast cancer cell lines with a constitutively activated extracellular-regulated kinase (ERK) pathway. Mol Cancer Ther. 2002;1:303–9.

Horibata S, Vo TV, Subramanian V, Thompson PR, Coonrod SA. Utilization of the soft agar colony formation assay to identify inhibitors of tumorigenicity in breast cancer cells. J Vis Exp. 2015;99:e52727.

Puck TT, Marcus PI, Cieciura SJ. Clonal growth of mammalian cells in vitro; growth characteristics of colonies from single HeLa cells with and without a feeder layer. J Exp Med. 1956;103:273–83.

Roberts AB, Anzano MA, Wakefield LM, Roche NS, Stern DF, Sporn MB. Type beta transforming growth factor: a bifunctional regulator of cellular growth. Proc Natl Acad Sci U S A. 1985;82:119–23.

Weaver VM, Lelièvre S, Lakins JN, Chrenek MA, Jones JC, Giancotti F, Werb Z, Bissell MJ. β4 integrin-dependent formation of polarized three-dimensional architecture confers resistance to apoptosis in normal and malignant mammary epithelium. Cancer Cell. 2002;2:205–16.

Xu Y, Afify SM, Du J, et al. The efficacy of PI3Kγ and EGFR inhibitors on the suppression of the characteristics of cancer stem cells. Sci Rep. 2022;12(1):347.

Chapter 17
Tumorigenic Potential of Cancer Stem Cells In Vivo

Abstract Cancer is a nightmare disease due to the tumor heterogeneity resulting in the requirement of multiple treatments. The cell plasticity will lead to tumor heterogeneity which means the presence of different cancer cells within the same tumor tissue. In heterogeneity, a hierarchy of cells exists as tumorigenic cancer cell populations and differentiated non-tumorigenic progenies. Cancer stem cell model explains this cellular hierarchy by assuming that cancer stem cells at the top of a hierarchical tree with the ability to self-renew and differentiate into all of the cancer cell lineages present within the tumor. In contrast to bulk tumor cells in old concepts, cancer stem cells are observed to have chemo−/radio-resistance demonstrating new tumor initiation and metastatic spread. Thus, cancer stem cells are responsible for recurrence and should be the true target of cancer therapy. To make this approach practical, malignant tumors developed from cancer stem cells would be a great help to understand the tumor heterogeneity.

In this chapter we show the method to develop malignant tumors from CSCs evaluating the histopathological analysis of the developed tumors. This method should renew the overview of cancer research which could help uncover the tumor heterogeneity.

Keywords Cancer stem cells · Cancer model · Tumor microenvironment

17.1 Introduction

Tumorigenesis is still controversial between the hierarchical models and the clonal model. Tumors are generally considered to initiate from any cells according to either the hierarchical or the clonal model. The clonal model has explained that mutation (s) in a cell and/or oncogenic factors affecting a cell may initiate tumor (Quail et al. 2012). On the other hand, the hierarchical model elucidates the heterogeneous population in a cancer tissue due to the cell plasticity (Afify and Seno 2019).

The CSC model belongs to the hierarchical model. According to the CSC model, tumors are formed as the result of CSC differentiation resulting in the heterogeneous progenies, which will lose stemness and become senescent to die at the end of differentiation. However, dedifferentiation may occur even in some differentiated

S. M. Afify, M. Seno, *Methods in Cancer Stem Cell Biology*,
https://doi.org/10.1007/978-981-99-1331-2_17

epithelial cells to transform into CSCs. In this case, the dedifferentiation can be called hierarchical or clonal (Chaffer and Weinberg 2015). CSCs and "tumor initiating cells (TICs)" describe the same cells that are predominant in tumor initiation (Afify et al. 2022a). Accordingly, the tumorigenic assay is a standard and well-accepted method to define CSCs (Chen et al. 2012; Calle et al. 2016; Nair et al. 2017; Afify et al. 2019; Afify et al. 2020; Minematsu et al. 2022; Du et al. 2020; Hassan et al. 2022; Afify et al. 2022b; Chen et al. 2022).

Xenografts in immunodeficient mice have successfully been demonstrated to evaluate the property of CSCs in vivo. Researchers have used this assay to determine the tumorigenic capacity of cells derived from a variety of transformants or isolated from malignant tumors. Fixation and paraffinization of the developed tumor tissues are the most important step to pathologically assess the CSC with different phenotypes and malignancy. Dissociation of primary tumors and the development of primary culture are also essential to analyze the alterations at the molecular level related to the aggressiveness and the metastatic potential. A combination of the pathological and cell biological data from the in vivo study will help us understand cancer initiation.

In the current chapter, we are trying to evaluate the tumorigenicity of CSCs developed from iPSCs in the conditions of inflammation.

17.2 Materials

17.2.1 Reagents

- Cancer stem cells (e.g., miPS-LLCcm cells) (Chen et al. 2012).
- Dulbecco's modified Eagle medium (DMEM, Sigma-Aldrich, St. Louis, MO, USA).
- Disodium hydrogen phosphate (Na2HPO4) (Sigma-Aldrich, cat. no. 255793).
- Fetal bovine serum (FBS; e.g., Gibco, Life Technologies, Massachusetts, USA).
- Penicillin/streptomycin mixed solution (Nacalai Tesque, Japan).
- Phosphate-buffered saline (Dulbecco's formula PBS) (Genesee Scientific, El Cajon, USA).
- Potassium chloride (KCl) (Sigma-Aldrich, cat. no. P9333).
- Potassium dihydrogen phosphate (KH2PO4) (Sigma-Aldrich, cat. no. P0662).
- MEM non-essential amino acids solution (×100) (Thermo Fisher Scientific, Waltham, MA, USA).
- L-Glutamine (Nacalai Tesque, Japan).
- Sodium chloride (NaCl) (Sigma-Aldrich, cat. no. 746398).
- 2-Mercaptoethanol (Sigma-Aldrich, St. Louis, USA).
- 2.25% Trypsin-EDTA (Atlanta Biologicals, Flowery Branch, USA).
- Gelatin, tissue culture grade (type B, 2% in H2O; Sigma-Aldrich, cat. no. G1393).

- Isoflurane (gas anesthesia system) DS Pharma Animal Health Co., Ltd. Osaka 541–0053, Japan).
- BALB/c-nu/nu, female, 4 weeks old, Charles River laboratories, Kanagawa, Japan.

17.2.2 Equipment

- Adjustable pipettes: P-20 (Gilson, cat. no. FA10003M), P-200 (Gilson, cat. no. FA10005M), and P-1000 (Gilson, cat. no. F123602M).
- Biosafety Cabinet (BSC-04IIA2) airtech.
- Safety Cabinet Class 2 (ESCO, model no. LA2-4A1).
- Cell culture incubator (CO2 at 5%, humidified at 37 °C) (Thermo Scientific cat. no. 311).
- CKX41 Inverted Microscope – Olympus Life Science (Japan).
- Cotton swabs (Sigma-Aldrich, cat. no. Z699365).
- Eppendorf 5415R Refrigerated Centrifuge (Eppendorf, Germany).
- Falcon® Conical Centrifuge Tubes (15 mL; BD Falcon, New York, NY, USA, Cat. No 352095).
- Hemocytometer (Sigma-Aldrich, cat. no. Z359629).
- Laser scanning confocal microscope, FV-1000 (Olympus, Tokyo, Japan).
- Liquid N2 storage tank.
- Microcentrifuge 1.5-mL tubes (Eppendorf, cat. no. 0030120086).
- Pipette tips 20 μl (Starlab, cat. no. S1120–1810), Pipette tips: 200 μl (Starlab, cat. no. S1120–8810), Pipette tips 1000 μl (Starlab, cat. no. S1120-1830).
- Sterile forceps.
- Syringes.
- Software/ImageJ – Computer Vision Online.
- Water bath (Thermo Scientific cat. no. 152–4101).
- Tissue culture–treated plate, 60 mm dish (TPP Techno Plastic Products AG Schaffhausen, Switzerland, Cat. No 93060).
- Tissue culture–treated plate, 100 mm dish (TPP Techno Plastic Products AG Schaffhausen, Switzerland, Cat. No 93100).
- Transwell inserts for 24-well plates (membrane 8.0 μm pores) (Corning, cat. no. 353097).
- 5/10/25-mL plastic disposable pipette.

17.2.3 *Reagent Preparation*

Stem Cell Medium

Mix the following per 500 mL		
Dulbecco's modified Eagle's medium-high glucose		412.5 mL
FBS		75 mL
Penicillin/streptomycin mixed solution	(50 U/mL)	2.5 mL
Gibco™ L-glutamine (200 mM)	(2 mM)	5 mL
MEM non-essential amino acids solution	(1 mM)	5 mL

PBS (pH 7.40) Preparation
Prepare PBS by mixing 137 mM NaCl, 10 mM Na2HPO4, 2.7 mM KCl, and 1.8 mM KH2PO4. Prepare the buffer in distilled water (dH2O), adjust the pH to 7.4 and autoclave it for 20 min.

0.1%(w/v) Gelatin
Dissolve 0.5 g of gelatin (from porcine skin) in 500 mL distilled water and autoclave. Store at 4 °C.

17.3 Methods

Our group developed a novel method to induce malignant tumors in nude mice (Chen et al. 2012; Afify et al. 2019) and this method has been extended to many cases of cancer study on CSCs derived from iPSCs (Chen et al. 2012; Calle et al. 2016; Nair et al. 2017; Afify et al. 2020). Since the preparation of malignant tumors is often required to study cancer development, the transplantation of CSCs will give you a good model (Fig. 17.1). All the following steps must be performed in a standard tissue culture hood in an aseptic way. Contamination and overgrowth should be avoided all through the experiment. Cells should be passaged at least 3 times before starting the experiment.

17.3.1 *Preparation of Gelatin-Coated Dishes*

1. Transfer 0.2% gelatin solution from 4 °C to the clean bench.
2. Cover the bottom of the dish with sufficient volume of 0.2% gelatin solution.
3. Incubate the dishes for 30 min at 37 °C in an incubator with 5% CO_2.

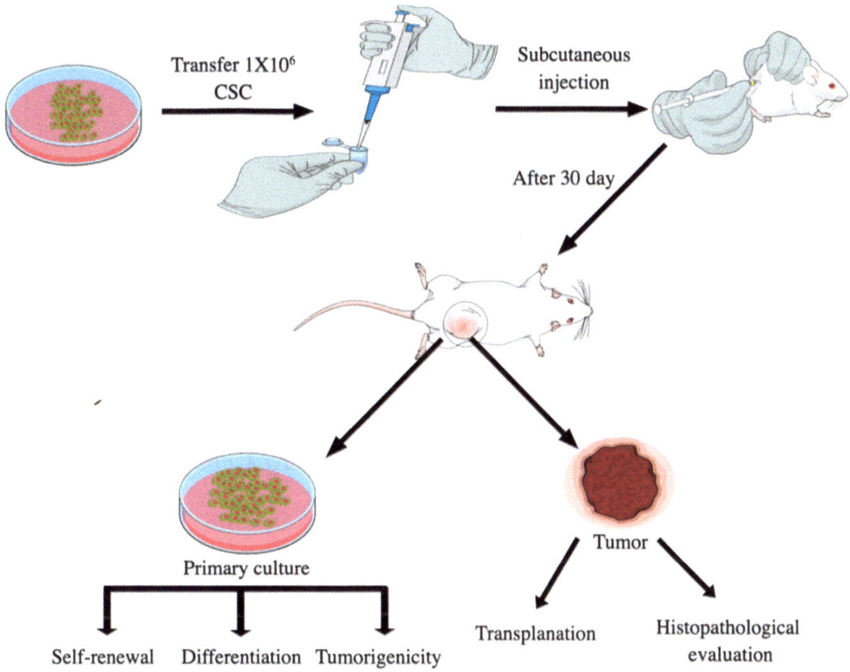

Fig. 17.1 Representative scheme for the preparation of tumors with CSCs induced from iPSCs

17.3.2 Cancer Stem Cell Thawing

1. Coat three 60-mm dishes with 0.1% gelatin at 37 °C in an incubator balanced with 5% CO_2 for at least 30 min.
2. Pre-warm the medium for cancer stem cell culture at 37 °C.
3. Transfer 5 mL of pre-warmed medium to a sterile 15-ml conical tube.
4. Pick up a vial of miPS-LLCcm cells from liq-N_2.
5. Thaw it in a 37 °C water bath incubating for 1–3 min.
6. Sterilize the vial with cotton in 70% ethanol.
7. Put the vial of cells inside of the cell culture hood.
8. Transfer the cells into a sterilized 15-mL conical tube.
9. Add 5 ml of pre-warmed medium prepared at Step 3 slowly into the tube dropping one by one with gentle mixing.
10. Centrifuge the tube at 200 × g for 5 min.
11. Remove the supernatant without disturbing the cell pellet.
12. Resuspend the cell pellet in 1 mL of pre-warmed medium.
13. Seed the cells on three gelatin-coated 60-mm dishes.
14. Incubate the cells at 37 °C in an incubator balanced with 5% CO_2.
15. Change medium at 2-day intervals observing the cell conditions every day. Note 1.
16. Observe the cells with GFP microscope (Fig. 17.2).

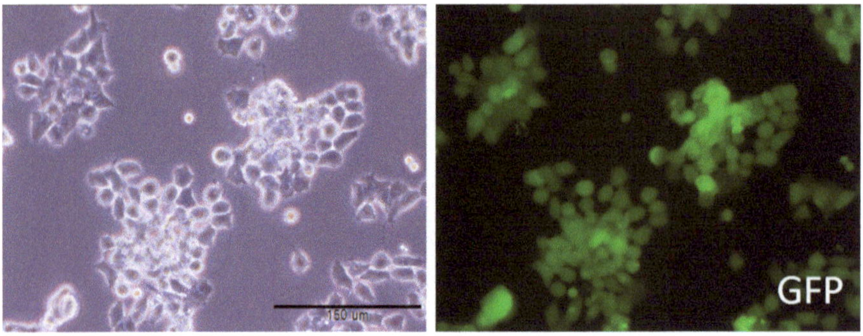

Fig. 17.2 Representative image for miPS-LLcm cells

17.3.3 Preparation and Injection of miPS-LLCcm Cells

1. Take cells at 70% confluent in a 60-mm dish.
2. Aspirate the medium.
3. Wash the cells with sterile PBS.
4. Trypsinize the cells.
5. Transfer the cells into a sterile 15-mL conical tube.
6. Centrifuge the tube at 200 × g for 5 min at 25 °C.
7. Aspirate the supernatant without disturbing the pellet.
8. Resuspend the pellet in 1 mL of sterile PBS.
9. Centrifuge the tube at 200 × g for 5 min at 25 °C.
10. Aspirate the supernatant without disturbing the pellet.
11. Suspend the cells in 1 mL of medium.
12. Take 20uL of the suspension and count the cells using a hemocytometer.
13. Suspend 1×10^6 cells in 100uL of sterile HBSS and then transfer the suspension into a 1.5 mL Eppendorf tube.
14. Keep the tube on ice until the mice get ready for injection. Note 2.
15. Inject the cells subcutaneously.
16. After 2 weeks a malignant tumor should be visible to develop. Note 3.
17. Take pictures of the mice with the tumor at any time.
18. Excise the tumor for histopathological studies and primary culture.
19. Excise the tumors then fix in 10% neutral formalin buffer solution for hematoxylin and eosin Y staining and immunohistochemical analysis.

17.3.4 Tumor Fixation, Paraffin Embedding, and Section Preparation

1. Fix tumor tissues in 10% formalin or 4% paraformaldehyde for 48 h at 4 °C. Note 4.

2. Rinse the tumor tissue with running tap water for 30 h.
3. Dehydrate the tumor tissue through as follows:

   I. In 30% ethanol for 30 min.
   II. In 60% ethanol for 30 min.
   III. In 70% ethanol for 30 min.
   IV. In 80% ethanol for 30 min.
   V. In 95% ethanol for 2 h twice.

4. In 100% ethanol for 2 h twice. Note 5.
5. Transfer the tissue into xylene to remove ethanol as follows.

   I. In ethanol: xylene (1:1, v/v) for 30 min.
   II. In 100% xylene for 30 min twice.

6. Immerse the tumor tissue in paraffin for 1 h with 3 repeats.
7. Embed the tissue in a paraffin block. Note 6.
8. Section the paraffin-embedded tissue block at 5 μm thickness on a microtome.
9. Float the sections in distilled water at 42 °C.
10. Transfer the sections onto a glass slide for immunohistochemistry.
11. Allow the slides to dry overnight and store slides at 25 °C until use.

17.3.5 Histochemical Observation with Hematoxylin and Eosin (H&E) Staining

1. Place slides of paraffin sections in a slide holder.
2. Clear the paraffin from the samples by immersing in xylene for 10 min.
3. Repeat Step 2 three times.
4. Hydrate the samples by transferring the slides in the following order:

   A. In 100% ethanol for 5 min twice.
   B. In 95% ethanol for 5 min.
   C. In 80% ethanol for 5 min.
   D. In 70% ethanol for 5 min.

5. Rinse the slides in running tap water for at least 5 min at 25 °C. Note 7.
6. Stain the slides in hematoxylin solution for 3 min at 25 °C.
7. Rinse the slides under running tap water at 25 °C for at least 10 min.
8. Stain the slides in 0.5% alcoholic eosin Y solution for 20 sec.
9. Dehydrate the samples by transferring the slides in the following order:

   A. In 95% ethanol for 3 min.
   B. In 95% ethanol for 3 min.
   C. In 100% ethanol for 3 min twice.

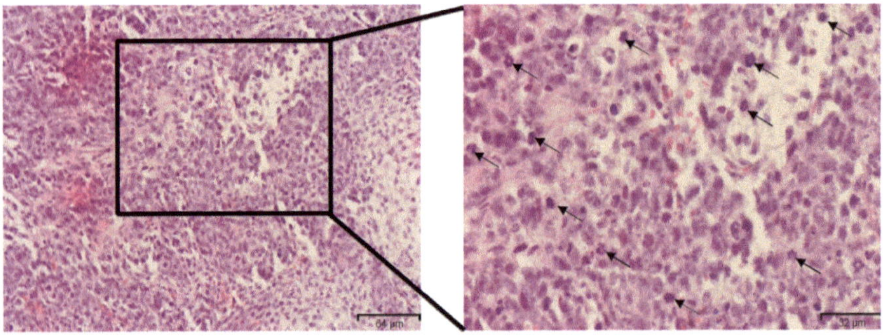

Fig. 17.3 Histopathological analysis of the malignant tumor derived from miPS-LLC cells. H&E staining of the malignant tumor, showing multiple pathologic mitotic figures (black arrow)

10. Clear the samples in three changes of xylene for 2 min per change.
11. Clear the samples by transferring the slides in the following order.

    A. In ethanol: xylene (50% v/v) for 10 min.
    B. In 100% xylene for 10 min twice.

12. Take out the slides from xylene and directly add a mounting medium. Note: Do not allow the slides to dry.
13. Observe the slides and take pictures under a microscope with magnifications of 20× and 40×. As shown the tumor derived from miPS-LLCcm cells exhibited malignancy with the mitotic figures (Fig. 17.3).

17.3.6 Immunohistochemistry for the Malignant Tumor

1. Take slides from Step 5 in the previous section.
2. Perform antigen retrieval to unmask the antigenic epitope as follows:

    I. Prepare antigen retrieval buffer (10 mM Sodium Citrate, pH 6.0).
    II. Fill a beaker with the buffer.
    III. Transfer the beaker into a pressure cooker with water. Note 8.
    IV. Place the pressure cooker on the hotplate and turn it on at full power.
    V. Put the slides from the tap water into the beaker for 10 min once boiling.
    VI. After 10 min turn off the pressure cooker.
    VII. Allow the slides to cool for 20 min.

3. Incubate the slides with PBS for 15 min and repeat the incubation 3 times.
4. Block endogenous peroxidase activity by incubating sections in 3% H_2O_2 in PBS for 20 min at 25 °C.

5. Rinse the slides with PBS for 15 min repeating 3 times.
6. Add 100 μL of blocking buffer (e.g., 10% fetal bovine serum in PBS) onto the sections of the slides.
7. Incubate the slides in a humidified chamber at 25 °C for 1 h.
8. Discard the blocking buffer from the slides.
9. Apply 200 μL of primary antibody appropriately diluted in antibody dilution buffer (e.g., 0.5% fetal bovine serum in PBS) to the sections on the slides.
10. Incubate the slides in a humidified chamber at 25 °C for 1 h.
11. Wash the slides with PBS for 15 min and repeat 3 times.
12. Apply 200 μl of biotinylated secondary antibody appropriately diluted in antibody dilution buffer to the slides.
13. Incubate the slides in a humidified dark chamber at 25 °C for 1 h.
14. Wash the slides with PBS for 15 min and repeat 3 times.
15. Place Avidin/Biotinylated Horseradish Peroxidase (HRP) Complex (ABC) kit on ice keeping it in the dark. Note 9.
16. Mix 2 drops of each A and B.
17. Incubate the mixture on ice 30 min keeping dark before use.
18. Add the mixture onto the section on the slide.
19. Keep dark for 30 min at 25 °C.
20. Wash the slides in PBS for 5 min and repeat twice.
21. Place the peroxidase substrate kit containing three different bottles of 3,3′--diaminobenzidine (DAB), hydrogen peroxide, and buffer pH 7.5 on ice.
22. Mix 2 drops of each hydrogen peroxide and buffer pH 7.5 and 4 drops of DAB.
23. Place the substrate mixture (final 0.05% DAB and 0.015% H_2O_2 in PBS) on ice. Note 10.
24. Apply 100 μL of substrate solution to the slides.
25. Allow less than 5 min until the desired intensity appears.
26. Wash slides with PBS for 5 min and repeat 3 times.
27. Counterstain the sections by immersing the sides in hematoxylin for 3 min.
28. Rinse the slides in running tap water for more than 15 min.
29. Dehydrate the tissue slides with ethanol as follows:

    I. In 95% ethanol for 3 min twice.
    II. In 100% ethanol for 3 min.
    III. In 100% ethanol for 5 min.

30. Remove ethanol and clear the slides and coverslips in xylene.
31. Repeat Step 28 three times with new xylene.
32. Mount the slides with coverslips using mounting solution. Note 11.
33. Observe the tissue sections with the developed color under a microscope e.g, Immunohistochemistry (IHC) staining for Ki67 is used to visualize Ki67 as a proliferation marker (Fig. 17.4).

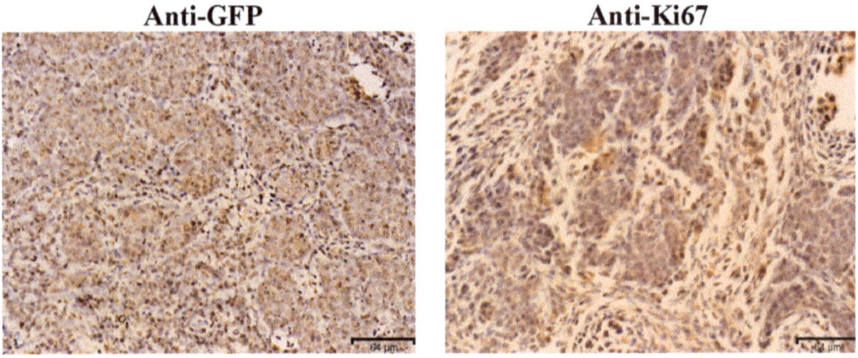

Fig. 17.4 Representative image for Immunohistochemistry (IHC) staining Ki67 showing cellular proliferation. Scale bars = 64 μm

17.4 Primary Culture from Malignant Tumor

1. After 30 days of transplantation or the size of the tumor is sufficient for the following analyses, anesthetize the mice with 2% isoflurane in an anesthesia chamber.
2. Sacrifice the mice and put in a clean bench with laminar flow.
3. Wipe the body with 70% ethanol.
4. Excise the tumor using sterilized scissors and transfer it into a sterile 100-mm dish.
5. Mince the tumor tissue into small pieces.
6. Transfer the pieces into a sterile 15-mL conical tube with 4 mL of dissociation buffer.
7. Incubate the tube at 37 °C for 6 h with shaking.
8. Terminate the digestion by adding 5 mL of DMEM containing 10% FBS.
9. Transfer the cell suspension into a new sterile 15-mL conical tube.
10. Centrifuge at 100 x*g* for 5 min at 25 °C.
11. Aspirate the supernatant without disturbing the pellet.
12. Resuspend the cell pellet in 5 mL of the same medium.
13. Centrifuge at 100 × *g* for 5 min at 25 °C.
14. Aspirate the supernatant without disturbing the pellet.
15. Suspend the cell pellet in an appropriate volume of iPSC complete medium without LIF.
16. Take 20 uL of the suspension and count the cells using a hemocytometer.
17. Seed the cells into sterile dishes at a density of 1×10^5 cells/cm^2. Note 12.
18. Aspirate the medium after 1 day.
19. Add a new fresh medium supplemented with puromycin to remove any cells from the host.
20. Observe the expression of GFP and cell morphology and then take photographs using Olympus IX81 microscope equipped with a light fluorescence device.

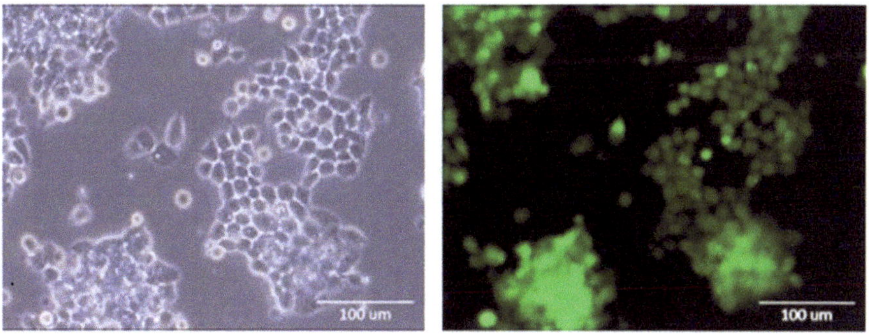

Fig. 17.5 Representative image for primary cultured cells derived from the malignant tumor developed from CSC. Scale bar: 100 μm

Primary culture of the malignant tumor showed two subpopulations of cells, GFP positive expressing cells and GFP negative cells (Fig. 17.5).

21. Subjecting the primary cultured cells to further experiments depends on your aim of study.

Notes

1. Do not leave the culture until the medium becomes yellow. Take photos at any time to memorize the morphology of the cells.
2. The cells should be alive for 1 h on ice.
3. Measure the size of the tumor every 2 days and monitor the growth.
4. Formalin is suspected as a carcinogen. Since it may cause irritation in the eye, skin, and respiratory tract, you should handle it in a hood.
5. You can incubate for up to 12 h but not more.
6. The paraffin tissue block can be stored at room temperature for years.
7. Slides in this step will be used in the next section.
8. Do not forget to put water in the cooker.
9. There are two bottles of A (Avidin) and B (Biotinylated HRP) in the kit.
10. DAB is suspected as a carcinogen. Prepare the solution just before use.
11. The mounted slides can be stored at 25 °C permanently.
12. Choose any size of dish depending on the total number of cells.

References

Afify SM, Hassan G, Osman A, Calle AS, Nawara HM, Zahra MH, El-Ghlban S, Mansour H, Alam MJ, Abu Quora HA, Du J, Seno A, Iwasaki Y, Seno M. Metastasis of cancer stem cells developed in the microenvironment of hepatocellular carcinoma. Bioengineering (Basel). 2019;6(3):73.

Afify SM, Hassan G, Seno A, Seno M. Cancer-inducing niche: the force of chronic inflammation. Br J Cancer. 2022a;127(2):193–201.

Afify SM, Hassan G, Yan T, Seno A, Seno M. Cancer stem cell initiation by tumor-derived extracellular vesicles. Methods Mol Biol. 2022b;2549:399–407.
Afify SM, Sanchez Calle A, Hassan G, Kumon K, Nawara HM, Zahra MH, Mansour HM, Khayrani AC, Alam MJ, Du J, Seno A, Iwasaki Y, Seno M. A novel model of liver cancer stem cells developed from induced pluripotent stem cells. Br J Cancer. 2020;122(9):1378–90.
Afify SM, Seno M. Conversion of stem cells to cancer stem cells: undercurrent of cancer initiation. Cancers (Basel). 2019;11(3):345.
Calle AS, Nair N, Oo AK, et al. A new PDAC mouse model originated from iPSCs-converted pancreatic cancer stem cells (CSCcm). Am J Cancer Res. 2016;6(12):2799–815.
Chaffer CL, Weinberg RA. How does multistep tumorigenesis really proceed? Cancer Discov. 2015;5:22–4.
Chen L, Kasai T, Li Y, et al. A model of cancer stem cells derived from mouse induced pluripotent stem cells. PLoS One. 2012;7(4):e33544.
Chen L, Liu Y, Xu Y, Afify SM, Gao A, Du J, Liu B, Fu X, Liu Y, Yan T, Zhu Z, Seno M. Up-regulation of Dsg2 conferred stem cells with malignancy through wnt/β-catenin signaling pathway. Exp Cell Res. 2022:113416.
Du J, Xu Y, Sasada S, Oo AKK, Hassan G, Mahmud H, Khayrani AC, Alam MJ, Kumon K, Uesaki R, Afify SM, Mansour HM, Nair N, Zahra MH, Seno A, Okada N, Chen L, Yan T, Seno M. Signaling inhibitors accelerate the conversion of mouse iPS cells into cancer stem cells in the tumor microenvironment. Sci Rep. 2020;10(1):9955.
Hassan G, Ohara T, Afify SM, Kumon K, Zahra MH, Fu X, Al Kadi M, Seno A, Salomon DS, Seno M. Different pancreatic cancer microenvironments convert iPSCs into cancer stem cells exhibiting distinct plasticity with altered gene expression of metabolic pathways. J Exp Clin Cancer Res. 2022;41(1):29.
Minematsu H, Afify SM, Sugihara Y, Hassan G, Zahra MH, Seno A, Adachi M, Seno M. Cancer stem cells induced by chronic stimulation with prostaglandin E2 exhibited constitutively activated PI3K axis. Sci Rep. 2022;12(1):15628.
Nair N, Calle AS, Zahra MH, Prieto-Vila M, Oo AKK, Hurley L, Vaidyanath A, Seno A, Masuda J, Iwasaki Y, Tanaka H, Kasai T, Seno M. A cancer stem cell model as the point of origin of cancer-associated fibroblasts in tumor microenvironment. Sci Rep. 2017;7(1):6838.
Quail DF, Taylor MJ, Postovit LM. Microenvironmental regulation of cancer stem cell phenotypes. Curr Stem Cell Res T. 2012;7:197–216.

Chapter 18
Development of Immunoliposomes Using Monoclonal Antibodies Targeting Cancer Stem Cells

Abstract A cell surface glycoprotein of hyaluronic acid (HA) receptor is known as CD44, which is often overexpressed in most types of tumors. CD44 has simultaneously been recognized as a cancer stem cell marker in many solid tumors. HA is considered to stimulate the signaling pathway of EGF receptor, through CD44 consequently leading to tumor cell growth, metastasis, and occasional recurrence. Recent studies have suggested that anti-CD44 monoclonal antibodies could target cells, neutralizing the interaction of CD44 with HA inhibiting the sequestrating signaling pathway. Thus, CD44 is one of the most favorable therapeutic targets to treat advanced stage of malignancies in the future.

In this chapter, we show an example to target cancer stem cells with anti-CD44 antibodies describing liposomes loaded with anti-cancer drugs. In this method, we demonstrate the good efficacy to target cancer stem cells with liposomes successfully conjugated with anti-CD44 antibodies to deliver the drug to the specific site.

Keywords Cancer stem cell · Active targeting · Hyaluronic acid (HA) · CD44, immunoliposomes

18.1 Introduction

The leading cause of global death is cancer (Momenimovahed et al. 2017). There is no doubt that ovarian cancer is one of the most common gynecologic cancers due to its high mortality rate. The number of death by ovarian cancer was 184,799 in 2018, representing 4.4% of all cancer-related deaths (Bray et al. 2018). It appears important to know the incidence and mortality of ovarian cancer in order to prepare for treatments and for prevention of complications.

Although the first priority is the histological diagnosis of the tumor including its grade in every case, systemic chemotherapy and surgical resection are currently the standard for ovarian cancer treatment (Kampan et al. 2015). Targeted drug delivery has been extensively investigated using liposomes. The first liposomes were developed by Alec Bangham in the early 1960s (Weissig 2017). Liposomes are spherical vesicles made up of lipid bilayers surrounding an aqueous core. The hydrophobic tails of the phospholipids are directed into the hydrophobic phase of

S. M. Afify, M. Seno, *Methods in Cancer Stem Cell Biology*,
https://doi.org/10.1007/978-981-99-1331-2_18

membrane bilayer, while their hydrophilic heads are directed to the aqueous phase. Through ligand–receptor interactions, active targeting could deliver the liposomes to target cells if the liposomes are displaying some ligands specific to the cell surface molecules. In cancer therapy, active targeting is considered to reduce off-target side effects because normal tissue may be affected by the toxicity without specific ligand–receptor interactions. The active-targeting can be designed by grafting peptides, proteins, monoclonal antibodies, aptamers, carbohydrates, and other small molecules on the surface of liposomes. Either the targeting moiety could be directly integrated into the lipid membrane or chemically conjugated to the radicals on the surface molecules so that, the liposomes can stably hold the targeting moiety until they bind the receptors (Steichen et al. 2013; Riaz et al. 2018; Jain and Jain 2018). Since monoclonal antibodies (mAbs) and their fragments (fragment antigen-binding (Fab)) recognizing targets are occasionally exploited as the ligand moiety in cancer drug delivery, the liposomes are so called "iImmunoliposomes (IL)" (Kutova et al. 2019).

Recently, our group evaluated the therapeutic efficacy of glycosylated paclitaxel (gPTX)-loaded LLs functionalized with anti-CD44 antibodies (gPTX-IL). We assessed the cytotoxicity of gPTX-IL against ovary cancer cell lines SK-OV-3 and OVK18 cells in vitro. The antitumor activity was also assessed in vivo being confirmed highly effective on cancer tissues with less side effects in normal tissues (Khayrani et al. 2019). Since we could successfully demonstrated good efficacy of the method of active targeting cancer stem cells using drug encapsulated ILs conjugated with anti-CD44 targeting, we describe this method in this chapter.

18.2 Materials

- Dipalmitoylphosphatidylcholine (DPPC), 1,2-distearoyl-sn-glycerol-3-phosphoethanolamine-N-[methoxy(polyethylene-glycol)-2000](mPEG–DSPE).
- 1,2-distearoyl-sn-glycerol-3-phosphoethanolamine-N-[maleimide (polyethylene glycol)-2000] (Mal–PEG–DSPE) (Tokyo, Japan).
- Cholesterol (Chol) Kanto Chemical Co., Inc. (Tokyo, Japan).
- Thiazolyl blue tetrazolium bromide (MTT).
- RPMI 1640 medium (Sigma-Aldrich, St Louis, MO, USA).
- Dulbecco's Modified Eagle MediDMEM; Thermo Fisher Scientific.
- Fetal bovine serum FBS (Gibco, Life Technologies, Massachusetts, USA).
- Penicillin/streptomycin mixed solution (Nacalai Tesque, Japan).
- L-glutamine (Nacalai Tesque, Japan).
- 2-Mercaptoethanol (Sigma-Aldrich, St. Louis, USA).
- Trypsin solution. 2.5% trypsin (17–160E; BioWhittaker, Walkersville, MD).
- Phosphate-buffered saline (PBS) without calcium or magnesium, (Genesee Scientific, El Cajon, USA).
- Hank's balanced salt solution (HBSS) Genesee Scientific, El Cajon, USA.
- 99.5% ethanol (FUJIFILM Wako Chemicals, Japan).

- BALB/c-nu/nu immunodeficient mice, female, four -weeks old (Charles River laboratories, kanagawa, Japan).
- Water bath with heating control.
- Cell counting instrument (e.g. hemocytometer, TC20 Automated Cell Counter (Bio-Rad) or Vi-Cell™ (Beckman)).
- Inverted microscope with bright field (CKX-51, Olympus, Japan).
- Syringe needles, 24G.
- Sterile cotton buds.
- Iris/eye scissors, straight.
- Operating scissors.
- Standard forceps, 12–13 cm length with, straight points.
- Sanyo MCO-19AIC(UV) CO2 incubator (Marshall Scientific, Hampton, WV, USA).
- Labculture® Class II, Type A2 Biological Safety Cabinets (Esco Life Sciences, Singapore).
- Olympus IX81 inverted microscope (Olympus, Tokyo, Japan).
- Laser scanning confocal microscope, FV-1000, Olympus, Tokyo, Japan.
- Tissue culture–treated plate, 60-mm dish (TPP Techno Plastic Products AG Schaffhausen, Switzerland, Cat. No 93060).
- Tissue culture–treated plate, 100-mm dish (TPP Techno Plastic Products AG Schaffhausen, Switzerland, Cat. No 93100).
- Falcon® 15-mL Conical Centrifuge Tubes (BD Falcon, New York, NY, USA, Cat. No 352095).
- Falcon® 50-mL Conical Centrifuge Tubes (BD Falcon, New York, NY, USA, Cat. No 352070).
- Liquid N2 storage tank.
- Microscope with bright field.
- 1.5-mL microcentrifuge tubes (Eppendorf, Germany).
- Microcentrifuge with refrigerator (5414R, Eppendorf, Germany).
- 100-mm dishes (tissue culture treated; Corning, cat. no. 353003).
- Anesthesia machine (Vet Tech Solutions).
- 5/10/25-mL plastic disposable sterile pipette (e.g. Falcon, Corning or Coaster).

18.3 Methods

18.3.1 Preparation of Anti-human CD44 mab

1. Pick up a vail of hybridoma Hermes-3 cells (HB-9480, ATCC, Manassas, VA) from liquid nitrogen using metal forceps.
2. Immerse the vial in a water bath at 37 °C without submerging the cap.
3. Pick up the vial from the water bath when the liquid defrosts half.
4. Sterilize the surface of the vial with 70% ethanol and place it in the hood.

5. Take off the cap and gently transfer the cells into a sterile 15-mL conical tube with a sterile Pasteur pipet.
6. Slowly add 1 ml of pre-warmed RPMI1640 or DMEM containing 10% FBS drop by drop.
7. Add further 3 ml of pre-warmed RPMI1640 or DMEM containing 10% FBS.
8. Centrifuge the tube at 200 × g for 5 min at 25 °C.
9. Aspirate the supernatant without disturbing the cell pellet and resuspend in 1 ml of pre-warmed RPMI1640 or DMEM containing 10% FBS then culture the cells at 37 ˚C in a CO2 incubator.
10. Count the cells when the cells becomes confluent.
11. Suspend 20 million cells in 50 mL of PFHM-II (Gibco, Grand Island, NY, USA) medium.
12. Transfer the cells into the tissue culture compartment of the bioreactor, miniPERM (SARSTEDT, Nümbrecht, Germany) then fill 350 mL of PFHM-II medium into the production module.
13. Connect the production module to the tissue culture compartment.
14. Rotate the bioreactor for 10 days at 37 °C in 5% CO2 on a rotating platform.
15. Collect the medium from the production module 50-mL conical tubes. If necessary to obtain more amount of the antibody, it is possible to add fresh medium for further incubation.
16. Centrifuge the medium at 150 × *g* for 5 min at 4 °C.
17. Transfer the supernatant to new 50-mL conical tubes.
18. Re-centrifuge at 10,000× for 5 min at 4 °C.
19. Pass supernatant through a 0.20 μm filter (Sartorius Stedim Biotech GmBH, Geottingen, Germany), Note 1.
20. Pass the supernatant through a protein A sepharose (bed volume 0.5 mL, GE Healthcare, Uppsala, Sweden) equilibrated with PBS.
21. Elute anti-human CD44 mAb using 0.1 M sodium-aceta buffer, pH 2.6 then collect the 5 mL of eluate by 10 fractions, i.e., 0.5 mL/ fraction.
22. Neutralize each 500 μL of each fraction with 10 μL of 2 M sodium phosphate buffer, pH 8.0.
23. Confirm the presence of anti-human CD44 mAb by Western blotting using polyclonal horse radish peroxidase labeled anti mouse IgG (DAKO, Glostrup, Denmark).
24. Determine the protein concentration by Bradford assay (BioRd Labs.) or a bicinchoninic acid (BCA) assay (Pierce Biotechnology, Rockford, IL, USA).

18.3.2 Preparation of Liposome Encapsulating gPTX (gPTX-L)

18.3.2.1 Preparation of gPTX-L

Liposomes composed of dipalmitoylphosphatidylcholine (DPPC) and cholesterol (Chol) at ratio 3:1 with 5 mol% 1,2-distearoyl-sn-glycerol-3-phosphoethanolamine-

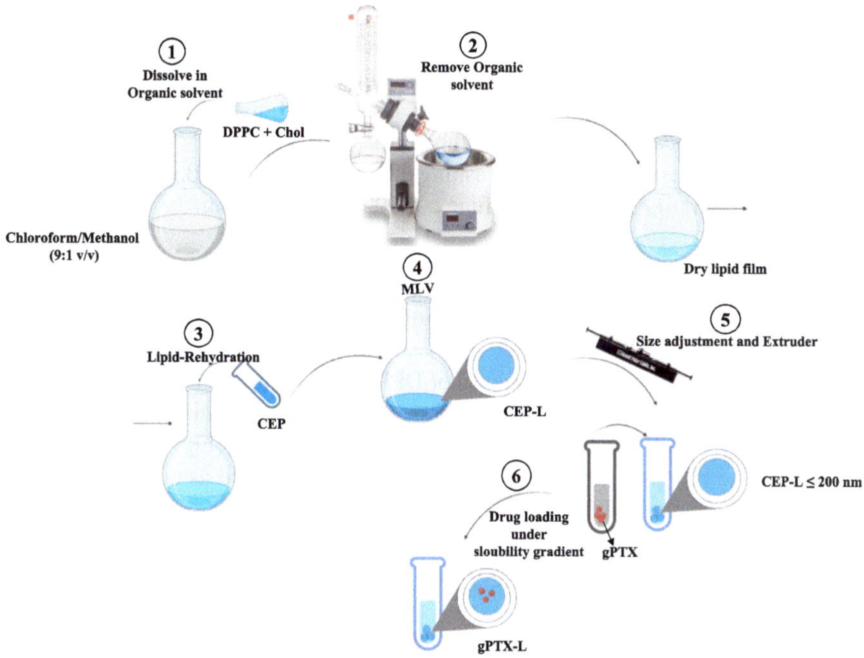

Fig. 18.1 The scheme of encapsulation of gPTX into liposome

N-[methoxy(polyethylene glycol)-2000] (mPEG–DSPE) were prepared by the thin-film hydration method (Fig. 18.1). gPTX is encapsulated into the liposome by the solubility gradient method as described below.

1. Dissolve DPPC and Chol with 5 mol% mPEG–DSPE in an organic solvent of chloroform/methanol (9:1 v/v) in an egg flask.
2. Connect the flask to a rotary evaporator (Eyela, Tokyo, Japan).
3. Maintain the rotary evaporator at 50 °C under vacuum with aspiration.
4. Keep rotation under vacuum overnight to remove organic solvent to make thin film of the lipid mixture.
5. Suspend the fully dehydrated lipid film in CEP (Chremophore, ethanol, PBS in 12:12:76 volume ratio) by vortexing at 60 °C, allowing the formation of multilamellar vesicles (MLVs).
6. Repeat freezing and thawing the MLVs for five cycles, Note 2.
7. Pass the MLVs 10 times through a single stack of one 100 nm polycarbonate membrane (Whatman, GE Healthcare, Carlsbad, CA, USA) using the Mini Extruder (Avanti Polar Lipids, Inc., Alabaster, AL, USA) to form small lamellar vesicles (SLVs), Note 3.
8. Replace the outer solvent CEP of the liposomes with PBS by ultrafiltration with a 100 K-membrane filter (Merck Millipore Ltd., Billerica, MA, USA) by five times of centrifuge at 5000 ×g for 20 min to prepare liposomes encapsulating CEP (CEP-L).

9. Add 1 mg/mL of gPTX solubilized in 40% ethylene glycol into the outer solution of CEP-L at 60 °C to allow the encapsulation of gPTX into CEP-L.
10. Concentrate the resultant gPTX-L to the volume equivalent to that at the end of Step 8, just before the addition of drug, by ultrafiltration.
11. Repeat the encapsulation process three times.
12. Remove the residual gPTX by washing the liposomes with PBS followed by five times of ultrafiltration by centrifuge at 5000 × g for 20 min.

18.3.2.2 Preparation of Immunoliposomes Containing gPTX (gPTX-IL)

1. Incubate CEP-L, which is composed of DPPC, Chol, mPEG–DSPE, with 0.5 mol % 1,2-distearoyl-sn-glycerol-3-phosphoethanolamine-N-[maleimide (polyethylene glycol)-2000] (Mal–PEG–DSPE) at 50 °C for 10 min (Fig. 18.2). Note 4.
2. Follow the encapsulation by the solubility gradient method as described in steps 9 to 12 in the above section 3.2.1.
3. Treat anti-human CD44 mAb with 2-iminothiolane (Sigma–Aldrich, St. Louis, MO, USA) at a molar ratio of 1:50 in 25 mM HEPES, pH 8.0 containing 140 mM NaCl, Note 5.
4. Incubate the mixture for 1 h at room temperature in the dark.
5. Remove unreacted 2-iminothiolane by gel filtration with a G25 PD-10 column (GE Healthcare, Uppsala, Sweden).

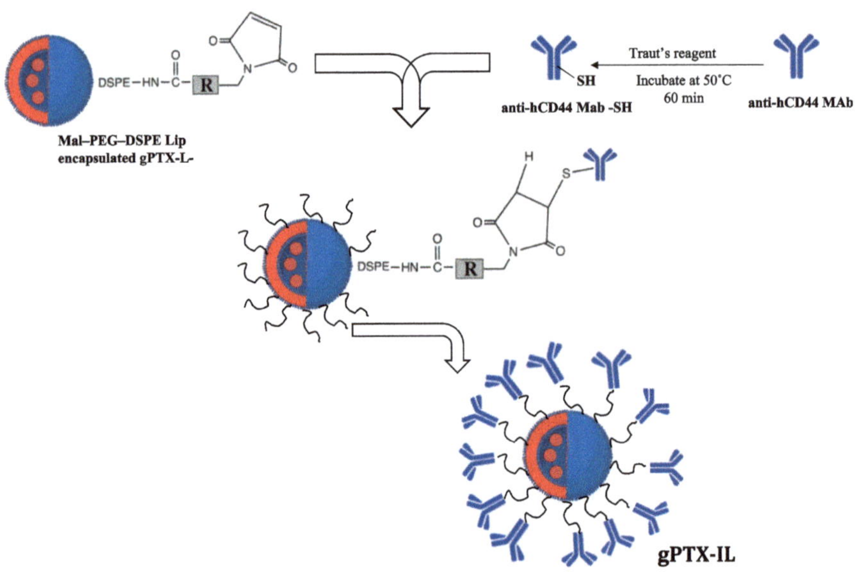

Fig. 18.2 Scheme for the preparation of gPTX-IL. R represents polyethylene-glycol moiety

6. Incubate the modified anti-human CD44 mAb with liposomes containing Mal–PEG–DSPE overnight at 4 °C.
7. Incubate liposomes with L-cysteine (0.5 mM final concentration) for 15 min at 25 °C, Note 6.
8. Remove residual L-cysteine by 5 times of centrifuge at 5000×g for 20 min of the solution through ultrafiltration membrane of 100 K-membrane filter Note 7.
9. Remove unreacted modified anti-human CD44 mAb by centrifuge at 6000 × g for 20 min at 4 °C for ultrafiltration with a 300 K membrane filter (Sartorius Stedim Biotech GmbH, Gottingen, Germany).

18.3.2.3 Evaluation of Antitumor Effects of Drugs In Vivo

1. Remove the medium from SK-OV-3 cells in a 100-mm dish.
2. Wash the cells with 5 mL of PBS.
3. Aspirate PBS.
4. Add 2 mL of trypsin.
5. Incubate for 5 min, or until cells have detached, at 37 °C.
6. Stop trypsinization by adding 10% FBS containing medium.
7. Transfer the cells in 15-mL sterile conical centrifuge tube.
8. Centrifuge the cells for 5 min at 1000 rpm and 25 °C.
9. Aspirate the supernatant without disturbing the pellet.
10. Wash the cells with 5 mL of sterile PBS.
11. Pellet the cells by centrifugation for 5 min at 1000 rpm and 25 °C.
12. Aspirate PBS, resuspend cells in fresh PBS.
13. Count the cells and prepare them at concentration 1×10^6 cells/100 μL.
14. Transfer cells to a sterile Eppendorf tube.
15. Inject 7.5×10^6 cells into the flank of an immune-deficient mouse (Fig. 18.3).
16. Measure the size of tumor every day with a Vernier calliper and calculate the volume by [length × (width)2]/2.
17. Start evaluation of the anti-tumor effect of each formulation when the tumor volume reaches 50–200 mm^3.
18. Assign mice to five groups ($n = 4$).

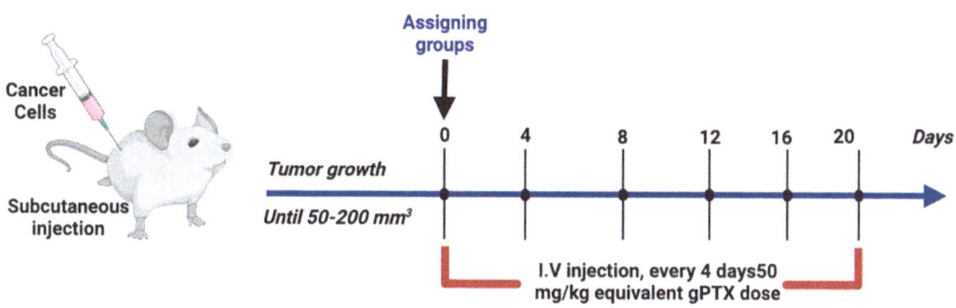

Fig. 18.3 Timeline of tumor growth and drug response in vivo

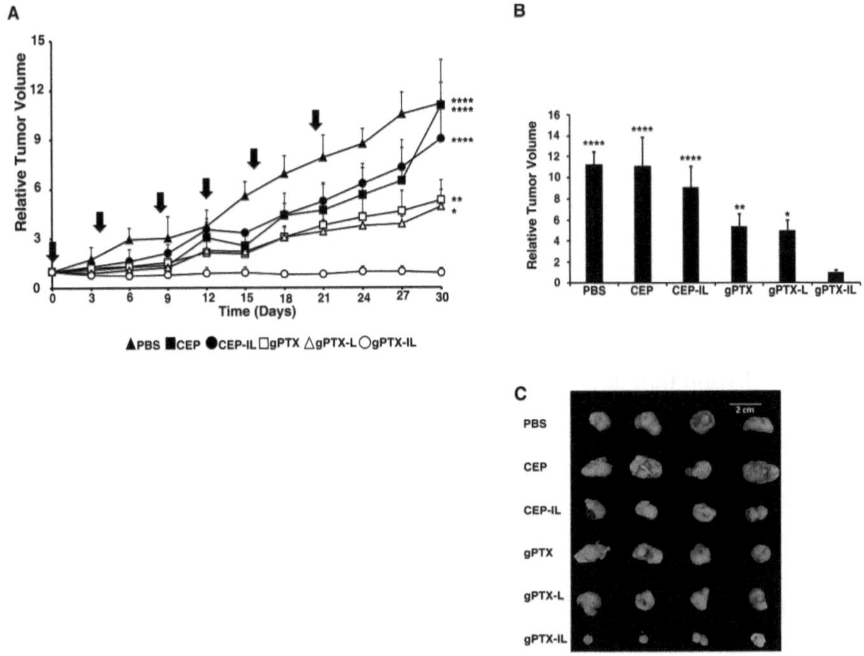

Fig. 18.4 Evaluation of the anti-tumor effect of gPTX-IL in vivo. gPTX-IL showed the most effective suppression of tumor growth (**a–c**). Statistical significance was determined by ANOVA and Dunnet multiple comparison test using the relative tumor volume of the gPTX-IL control group as a criterion for statistical significance, (*) $p < 0.05$; (**) $p < 0.01$; (****) $p < 0.001$. Data are expressed as the mean with $\pm$ SD where $n = 4$. The figure is copied from our original paper: (Khayrani et al. 2019)

A. Group 1 for PBS.
B. Group 2 for CEP.
C. Group 3 for CEP-IL.
D. Group 4 for naked gPTX.
E. Group 5 for gPTX-L and group 6 for gPTX-IL.

19. Inject 50 mg of gPTX-equivalent per kg body weight six times via tail vein at the intervals of 4 days.
20. Measure tumor volumes and body weights at 3- or 4-day intervals (Fig. 18.4).
21. Embed all organs including liver, kidney, and spleen sections.
22. Stain slides with hematoxylin (Sigma Aldrich, St. Louis, MO, USA) and 0.5% Eosin Y (Wako, Osaka, Japan) (HE) for histological analysis then visualized under a FSX100 inverted microscope (Olympus, Tokyo, Japan) (Fig. 18.5).

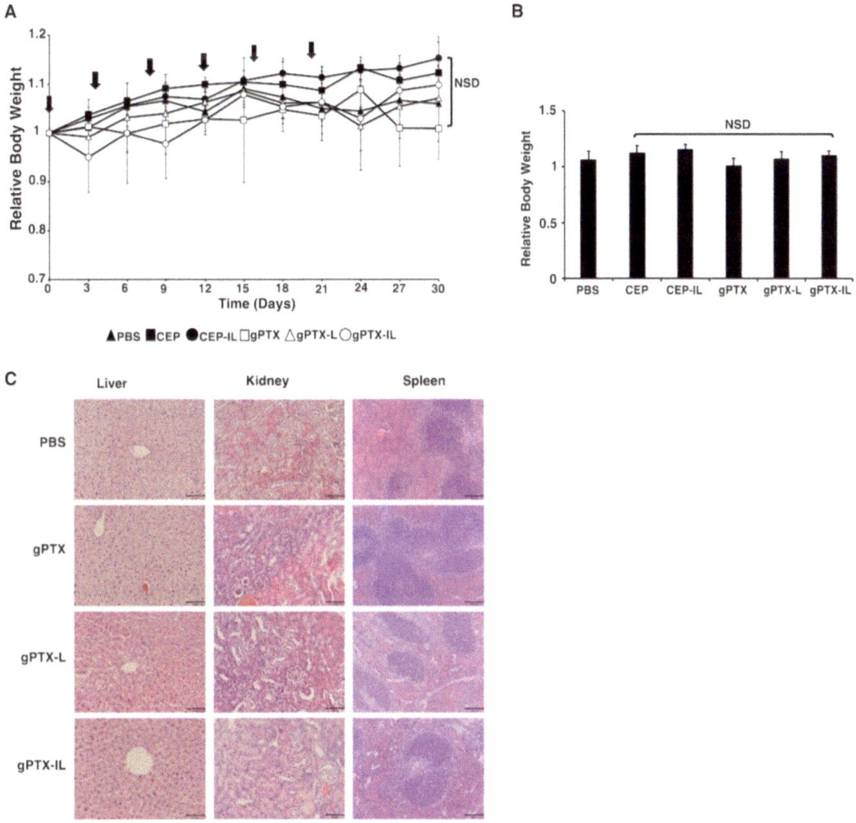

Fig. 18.5 Evaluation of the side effects of each formulation of gPTX. gPTX- IL (open circle with line), gPTX-L (open triangle), naked gPTX (open square), CEP-IL (closed circle), CEP (closed square), or PBS (cross) was intravenously injected at days 0, 4, 8, 12, 16, and 20 (indicated by vertical arrows). (**a**) Change of body weight of mice bearing tumors. (**b**) Relative body weight at day 30. The statistical significance in mean values of more than two groups was determined using one-way analysis of variance (ANOVA) and Dunnet multiple comparison test using relative body weight of PBS treatment as a control, NSD, no significant difference. (**c**) H&E staining of some vital organs of the drug-treated animal groups. Each scale bar of liver and kidney section shows 64 μm and that of spleen section shows 129 μm. The treatment with gPTX-IL showed no significance side effects when compared with the other formulation. The figure is copied from our original paper: (Khayrani et al. 2019)

Conclusion:

Pathological observations of liver, kidney, and spleen showed that gPTX-IL did not cause significant damage to the tissues during the experiment while naked drug and non-targeting liposome appeared to damage the tissues. In naked gPTX treatment, liver showed a cytoplasmic vacuolation. Kidney exhibited atrophied glomeruli and necrotic areas, and the spleen had enlargement of lymphoid follicles of white pulp. In gPTX-L treatment, elongated trabeculae were found in the spleen.

Meanwhile, no damage was found in any of those tissues in gPTX-IL treatment. The tumor growth appeared completely suppressed by the administration of gPTX-IL. As a result, only gPTX-IL was concluded to have the most excellent anti-tumor activity with less side effects among the formation assessed in the study.

Notes

1. This step showed be performed to completely remove cell debris.
2. A single freeze-thaw cycle consisted of freezing at -196 °C liquid nitrogen for 1 min and thawing at 55 °C water bath for 1 min.
3. The extruder was kept warm at 55 °C on a hot plate prior to extrusion.
4. This step is essential for introducing maleimide functional groups into liposome to conjugate antibodies.
5. This step is essential for introducing SH groups into anti-human CD44 mAb, which require to immobilize antibody on the surface of the liposomes.
6. This step is essential to block free maleimide groups.
7. This step is critical to remove the residual L-cysteine.

Acknowledgments It is our pleasure to acknowledge Dr. Apriliana Cahya Khayrani for publishing the original article in the International Journal of Molecular Sciences.

References

Bray F, Ferlay J, Soerjomataram I, Siegel RL, Torre LA, Jemal A. Global cancer statistics 2018: GLOBOCAN estimates of incidence and mortality worldwide for 36 cancers in 185 countries. CA Cancer J Clin. 2018;68(6):394–424.

Jain A, Jain SK. Advances in tumor targeted liposomes. Curr Mol Med. 2018;18:44–57.

Kampan NC, Madondo MT, McNally OM, Quinn M, Plebanski M. Paclitaxel and its evolving role in the management of ovarian cancer. Biomed Res Int. 2015;2015:413076.

Khayrani AC, Mahmud H, Oo AKK, Zahra MH, Oze M, Du J, Alam MJ, Afify SM, Quora HAA, Shigehiro T, Calle AS, Okada N, Seno A, Fujita K, Hamada H, Seno Y, Mandai T, Seno M. Targeting ovarian cancer cells overexpressing CD44 with immunoliposomes encapsulating glycosylated paclitaxel. Int J Mol Sci. 2019;20(1042):10.

Kutova OM, Guryev EL, Sokolova EA, Alzeibak R, Balalaeva IV. Targeted delivery to tumors: multidirectional strategies to improve treatment efficiency. Cancers. 2019;11:68. https://doi.org/10.3390/cancers11010068.

Momenimovahed Z, Ghoncheh M, Pakzad R, Hasanpour H, Salehiniya H. Incidence and mortality of uterine cancer and relationship with human development index in the world. Cukurova Med J. 2017;42(2):233–40.

Riaz MKA, Riaz MKA, Zhang X, Lin C, Wong KH, Chen X, Zhang G, Lu A, Yang Z. Surface functionalization and targeting strategies of liposomes in solid tumor therapy: a review. Int J Mol Sci. 2018;19:195.

Steichen SD, Caldorera-Moore M, Peppas NA. A review of current nanoparticle and targeting moieties for the delivery of cancer therapeutics. Eur J Pharm Sci. 2013;48:416–27. https://doi.org/10.1016/j.ejps.2012.12.006.

Weissig V. Liposomes came first: the early history of liposomology. Methods Mol Biol. 2017;1–15 https://doi.org/10.1007/978-1-4939-6591-5_1.

Chapter 19
In Vitro Evaluation of Anti-Cancer Stem Cell Drugs

Abstract Cancer cells will exhibit abnormal growth due to the disruption of the control of gene expression genome wide. The altering gene expression leads to escape from the cell cycle checkpoints directing to immortality with continuous cell proliferation in place of cell death. As the results showed, the cells contain cancer stem cells, which are considered resistant to anti-cancer drugs and/or radiation, even though cancer is a complex disease.

However, the search for anti-cancer stem cell drugs is an essential process, because this approach has scarcely been done before. The process will require a number of in vitro and in vivo studies including and clinical trials. To begin with, during the drug development process, in vitro assays serve as an initial platform for discovering potential drugs. Thus, it is logical to utilize them in the screening of anti-cancer stem cell drugs, as it offers a more viable solution for targeting the root cause of cancer. In the present chapter, we describe an assay procedure commonly utilized in vitro and techniques available to examine the viability, proliferation, and apoptosis of cells.

Keywords MTT · Cancer stem cells · Drug response

19.1 Introduction

When a new anti-cancer drug either from natural or synthetic source is under development, its cytotoxic effect on cancer cells as well as its safety to the host is required to be evaluated. This is well known as the cell viability test. This test may vary depending on the measurements based on various cellular functions such as enzyme activity, cell membrane permeability, cell adherence, ATP synthesis, co-enzyme production, and nucleotide uptake activity. Among them, 3-(4,5-dimethylthiazol-2-yl)-2,5-diphenyltetrazolium bromide (MTT), which depend on mitochondrial NAD(P)H oxidoreductase activity of viable cells, is one of the most frequently used reagents. Therefore, MTT assay is available to evaluate cell viability not only for cytotoxicity but also proliferation.

Since MTT (yellow salt) yields water-insoluble formazan crystal (purple) after the enzyme reaction and absorbance of formazan resolved in acid is measured, there

S. M. Afify, M. Seno, *Methods in Cancer Stem Cell Biology*,
https://doi.org/10.1007/978-981-99-1331-2_19

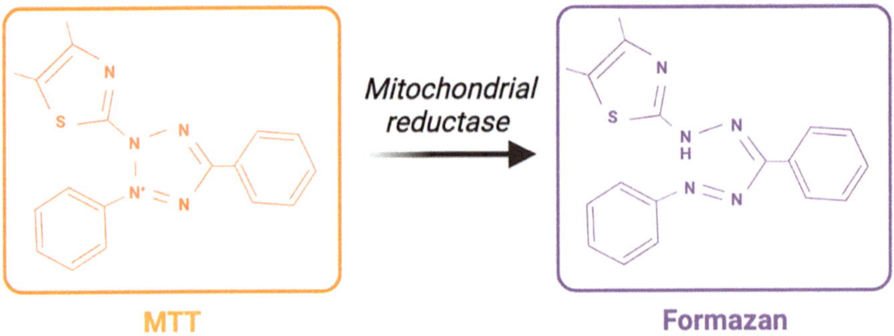

Fig. 19.1 In live cells, mitochondrial reductase reduces MTT into insoluble formazan

will be no insoluble yields in the absence of cells making low background absorbance (Fig. 19.1) (Vistica et al. 1991; Maehara et al. 1987; Slater et al. 1963; Berridge and Tan 1993; Markossian et al. 2004).

Since the linear relationship between the number of metabolically active cells and the amount of formed formazan was recognized, the rate of cell death or proliferation was found quantified by measuring the absorbance in the range of 500–600 nm derived from solubilized formazan. High purple color intensity denotes higher cell number while low signifies small cell number. The MTT assay has commonly been applied to a method suitable for high-throughput drug screening. This colorimetric assay was first designed by Mosmann et al. (Mosmann 1983).

Careful attention should be paid in the use of MTT because non-enzymatic reactions with reducing molecules like ascorbic acid, glutathione, or coenzyme A can interact with MTT forming the formazan, which may produce false results (Riss et al. 2004).

19.2 Materials

19.2.1 Reagent

- Dulbecco's modified Eagle medium (DMEM) (Fujifilm-Wako, USA).
- 3-(4,5-Dimethylthiazol-2-yl)-2,5-Diphenyltetrazolium Bromide (MTT) (Sigma, USA).
- Ethanol (EtOH) 95%.
- Compounds of drug candidates (inhibitors).
- Dimethyl sulfoxide (DMSO).
- Fetal bovine serum (FBS).
- Sterile distilled or ultrapure water (dH_2O/UPW).

19.2.2 *Equipment*

- Laminar flow clean bench/cabinet (standard: "biological hazard").
- Microtiter plate reader with grating or sufficient wavelength filters.
- Incubator: 37 ± 1 °C, 90 ± 5% humidity, 5.0 ± 1% CO_2/air.
- Inverted microscope with objective lenses ×4, ×10, ×20, and ×40.
- Sterile conical tubes (15, 50 mL).
- Sterile disposable serological pipettes (1, 5, 10 mL).
- Sterile micropipette tips (2, 10, 20, 200 microL).
- 96-well flat-bottom tissue culture microtiter plates.
- Hemocytometer.
- Micropipettes (2, 10, 20, 200, 1000 microL).
- Multi-channel micropipettes (50, 200 microL).
- Benchtop centrifuge.
- Autoclavable plastic microtubes (0.2, 0.5, 1.5 mL).
- Sterile cryo-tubes.

19.2.3 *Reagent Preparation*

19.2.3.1 MTT Solution

Dissolve MTT in Dulbecco's phosphate-buffered saline, pH = 7.4 (DPBS) to 5 mg/ml followed by vortexing or sonication. Sterilize the MTT solution passing through a 0.2-μm syringe-top-filter into a sterile conical tube. Wrap the tube with aluminum foil to protect from light. MTT solution should be kept at −20 °C (stable for at least 6 months) but at 4 °C with protection from light for frequent use.

19.2.3.2 Phosphate-Buffered Saline Solution (PBS)

Dissolve 8 g NaCl, 0.2 g potassium chloride (KCl), 1.44 g di-sodium monohydrate phosphate (Na_2HPO_4, $2H_2O$), 0.2 g monopotassium dihydrate phosphate (KH_2PO_4) in 1000 ml of dH_2O/UPW. Adjust pH to 7.4 using sodium hydroxide (NaOH) or hydrogen chloride (HCl).

19.2.3.3 Trypan Blue Solution

Dissolve 4 g of trypan blue in 1000 ml of 0.9% NaCl solution. Pass through a 0.45-μ m filter to remove undissolved trypan blue crystals.

19.3 Methods

In the current chapter, we evaluate the screening procedure of anti-cancer drugs targeting CSC using MTT. First, CSCs are seeded in a 96-well plate at 5000 cells/well leaving some wells empty for the reference. After 24 h, the cells are treated with various concentrations of the desired drug (Fig. 19.2). After at least further 24 hours, 20 μL of 5 mg/mL MTT in PBS (MTT working solution) is added to each well followed by the addition of 10% sodium dodecyl sulphate (SDS) in 5% HCl to dissolve the formed formazan. Then the absorbance of the colored solution is usually measured at between 500 and 600 nm by a microplate reader. All solutions, glassware, pipettes, etc., have to be sterile and all procedures should be carried out under aseptic conditions and in the sterile environment of a laminar flow cabinet (biological hazard standard).

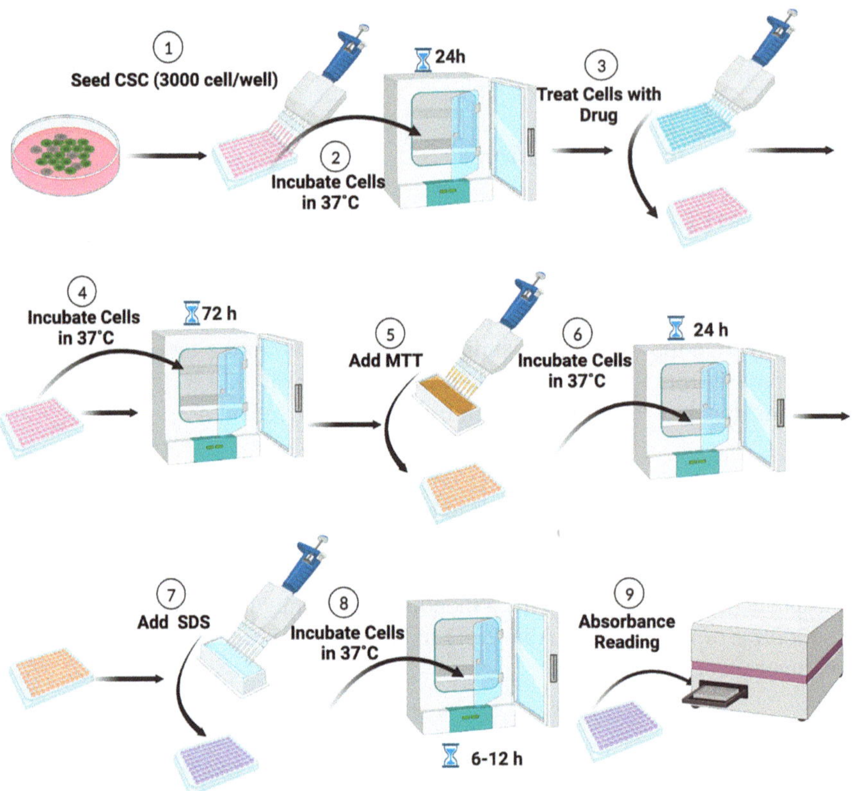

Fig. 19.2 This diagram illustrates the use of 3-(4,5-dimethylthiazol-2-yl)-2,5-diphenyl-2H-tetrazolium bromide) (MTT) as a colorimetric assay for cell viability

19.3.1 Cancer Stem Cells Culture

1. Pick up the vail of CSCs from liquid N_2.
2. Immerse the vial in a 37 °C water bath. Note 1.
3. Swire the vial gently.
4. Pick up the vial from the water bath.
5. Spray the outside of the vial with 70% ethanol.
6. Place it in a clean bench.
7. Pipette up the thawed cells gently into a sterile 15-mL conical tube.
8. Add 3 mL of pre-warmed MEF medium to cells in the 15-mL conical tube.
9. Rinse the vial with 1 mL of pre-warmed MEF medium and transfer to the same 15-mL conical tube with cells.
10. Centrifuge the cells at 200 × g for 5 min.
11. Aspirate the supernatant without disturbing the cell pellets.
12. Suspend the cells in 2 mL of pre-warmed MEF medium.
13. Seed the cells in two 60-mm dishes coated with 0.2% gelatin.
14. Incubate at 37 °C in a 5% CO_2 incubator.
15. Make passage to new dishes when the cells become 70% confluent (Fig. 19.3).

19.3.2 Cell Plating for MTT

1. Coat 96-well plates with 100 μl of 0.2% gelatin per well.
2. Incubate 30 min at 37 °C in a 5% CO_2 incubator. Note 2.
3. Aspirate the medium when the cells become 70% confluent in 60-mm dishes.
4. Wash the cell layer with 5 ml of PBS.
5. Aspirate the PBS.
6. Add 3 mL of 0.25% trypsin and wait until the cells are detached.
7. Stop the reaction and suspend the cells by adding 10% FBS medium.

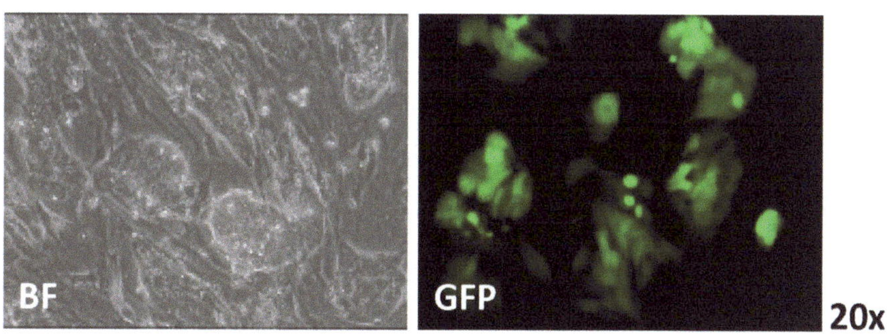

Fig. 19.3 Representative image for primary culture of CSC-derived malignant tumor in liver (Afify et al. 2020)

8. Transfer the cell suspension to a sterile 15-mL conical tube.
9. Centrifuge the tube at 1000 rpm for 5 min at 25 °C.
10. Discard the supernatant.
11. Add 10 mL of fresh medium.
12. Take 50 μL of the cell suspension and mix with 50 μL of 0.4% trypan blue solution.
13. Fill the chamber of a hemocytometer with 10–20 μL of the mixture.
14. Count the total number of cells, calculate the cell density, and adjust the density to 5×10^4 cells/mL with medium.
15. Bring the gelatin-coated 96-well plate from Step 2.
16. Aspirate the gelatin solution in each well.
17. Seed 5000 cells well in a 96-well plate. Note 3.
18. Incubate at 37 °C in a 5% CO_2 incubator overnight to allow the cells to attach to the wells.

19.3.3 Drug Preparation

1. After 24 h, treat the cells with various concentrations of the drug candidates. Final volume should be 100 μL per well. Note 4.
2. Incubate the cells for 72 h at 37 °C in a 5% CO_2 incubator. Note: The incubation period of 72 h may vary from 24 to 72 h depending on the evaluation (Fig. 19.4).

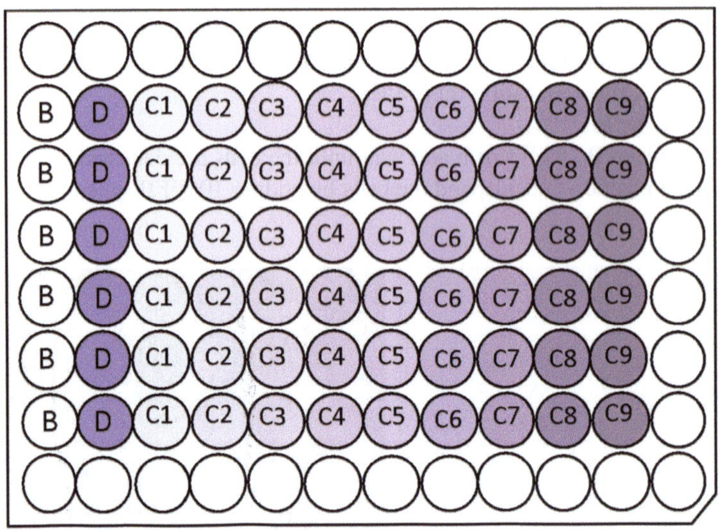

Fig. 19.4 Plate layout for cell exposure. B blank (no cells), D, cells treated with DMSO as control, C1–C9 test concentration in descending order

19.3.4 Formazan Formation

1. Add 20 μL of MTT working solution to each well after the appropriate incubation period of 24–72 h. Note: Include the wells with no cells as the reference. All steps should be done aseptically.
2. Incubate the plate for 4 h at 37 °C in a 5% CO_2 incubator.
3. Add 100 μL of 10% SDS/5% HCl per well.
4. Incubate overnight at 37 °C in a CO_2 incubator.
5. Measure the absorbance of each well at 570 nm using a microplate reader. Note 5.

19.3.5 Data Analysis

1. Average the triplicate reading for each sample.
2. Subtract the culture medium background from each reading of the sample to obtain the corrected absorbance. Note 6.
3. Calculate the IC_{50} for each drug (Fig. 19.5).

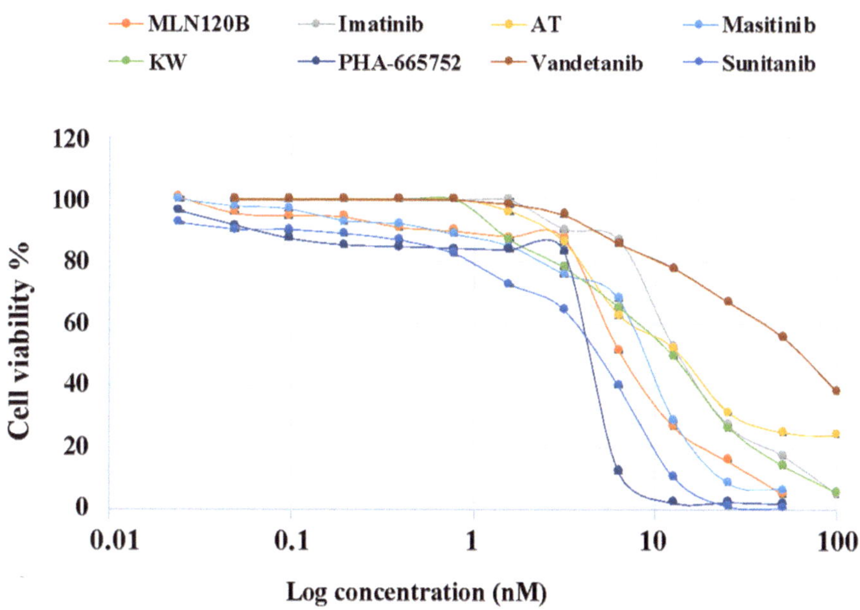

Fig. 19.5 Dose–response growth curves of CSCs treated with eight drugs, where $IC_{50} \leq 10$ μM in all developed CSCs in A. Dots represent the mean of the triplicates. Each error bar represents SD. miPS-huh7cm P cells showed high sensitivity to PHA-665752, Sunitinib, MLN120B, and Masitinib, while the cells showed resistance to Vandetanib, Imatinib, AT, and KW inhibitors

Notes

1. Avoid submerging the cap.
2. This plate is used at Step 15.
3. Leave eight wells empty for blank controls.
4. You must be very careful not to dispose of the cells when you remove media. If you fail, the results get unreliable.
5. Use a wavelength of 650 nm as reference to determine the background noise caused by undissolved particles and cell debris.
6. The absorbance is proportional to cell number where there is a linearity.

References

Afify SM, Sanchez Calle A, Hassan G, et al. A novel model of liver cancer stem cells developed from induced pluripotent stem cells. Br J Cancer. 2020;122:1378–90.

Berridge MV, Tan AS. Characterization of the cellular reduction of 3-(4,5-dimethylthiazol-2-yl)-2,5-diphenyltetrazolium bromide (MTT): subcellular localization, substrate dependence, and involvement of mitochondrial electron transport in MTT reduction. Arch Biochem Biophys. 1993;303:474.

Maehara Y, et al. The ATP assay is more sensitive than the succinate dehydrogenase inhibition test for predicting cell viability. Eur J Cancer Clin Oncol. 1987;23:273–6.

Markossian S, Grossman A, Brimacombe K, Arkin M, Auld D, Austin C, Baell J, Chung TDY, Coussens NP, Dahlin JL, Devanarayan V, Foley TL, Glicksman M, Haas JV, Hall MD, Hoare S, Inglese J, Iversen PW, Kales SC, Lal-Nag M, Li Z, McGee J, McManus O, Riss T, Saradjian P, Sittampalam GS, Tarselli M, Trask OJ Jr, Wang Y, Weidner JR, Wildey MJ, Wilson K, Xia M, Xu X, editors. Assay guidance manual [Internet]. Bethesda (MD): Eli Lilly & Company and the National Center for Advancing Translational Sciences; 2004.

Mosmann T. Rapid colorimetric assay for cellular growth and survival: application to proliferation and cytotoxicity assays. J Immunol Methods. 1983;65(1-2):55–63.

Riss TL, Moravec RA, Niles AL, et al. Cell Viability Assays. 2013 May 1 [Updated 2016 Jul 1]. In: Markossian S, Grossman A, Brimacombe K, et al., editors. Assay Guidance Manual [Internet]. Bethesda (MD): Eli Lilly & Company and the National Center for Advancing Translational Sciences; 2004.

Slater TF, Sawyer B, Straeuli U. Studies on succinate-tetrazolium reductase systems: III. Points of coupling of four different tetrazolium salts III. Points of coupling of four different tetrazolium salts. Biochim Biophys Acta. 1963;77:383–93.

Vistica DT, et al. Tetrazolium-based assays for cellular viability: a critical examination of selected parameters affecting formazan production. Cancer Res. 1991;51:2515–20.

Chapter 20
In Vitro Tumoroid Model Using Cancer Stem Cells

Abstract In vitro anti-tumor drug screenings are mainly performed by detecting the reaction of tumor cells in two-dimensional (2D) monolayer culture. Due to the convenient availability, 2D culture techniques are employed prior to in vivo in most of the process of evaluations of the efficacy of anti-cancer drugs and their candidates. However, the efficacy in vitro apparently has become weak in feasibility because of the increase of clinical trials resulting in failure. This has at least partially been attributed to the fact that cells grown in 2D culture lack the complex three-dimensional (3D) tissue architecture that realizes cell–cell or cell–extracellular matrix (ECM) interactions which can mimic tumors in real patients. 3D cancer models reflecting the pathophysiology and the resistance to radiation or drugs in vivo offer vital approaches for the development of novel theranostic drugs and systems by high-throughput drug screening. On the other hand, a subpopulation of stem-like cells within the tumor tissue so-called cancer stem cells (CSCs) are currently considered responsible for the tumor progression and metastasis. Therefore, 3D models with CSCs should be feasible for the screening of drugs and the development of personalized medicine. In the current chapter, we describe a detailed protocol to develop a CSC-based organoid as the 3D model. In this protocol we are showing two effective ways to develop 3D models using suspension and/or embedded methods.

Keywords Cancer stem cells · Three-dimensional · Tumoroid

20.1 Introduction

In the last decade, translational cancer research has witnessed a revolution with the development of methods that enable the reproducible derivation, maintenance, and biobanking of primary human normal and cancer tissues. These primary cell cultures will provide the models of human cancer demonstrating the heterogeneity helping understand cancer biology and develop effective treatments. The advanced models mimicking human cancer will be more helpful to develop personalized treatment with durable effects but less side effects.

S. M. Afify, M. Seno, *Methods in Cancer Stem Cell Biology*,
https://doi.org/10.1007/978-981-99-1331-2_20

Traditionally, the 2D cell culture techniques have been widely used in cancer research while these techniques cannot mimic in vivo conditions and processes, such as extracellular matrices with cell-to-cell adhesion molecules inducing specific signaling pathways (Baker and Chen 2012; Horning et al. 2008; Yamada and Cukierman 2007).

The 3D stem-cell derived cultures, so-called organoids, support the propagation of phenotypic, genetic, and transcriptomic characteristics of the original tissue retaining the indefinite ability of stem cells to self-renew and to undergo multilineage differentiation. This 3D method will facilitate cancer development in vitro mimicking the tumor microenvironment in vivo that controls the cellular phenotypes (Yamada and Cukierman 2007). In this context, various 3D culture models exploiting tissue engineering principles have been developed (Szot et al. 2011; Zou et al. 2010; Kievit et al. 2010; Hsiao et al. 2009). In this chapter, we try to develop a 3D model of CSCs, which have been converted from human iPSCs (Seno et al. 2016) under the inflammatory microenvironment derived from cancer cell lines.

20.2 Materials

20.2.1 Reagents

- Cancer stem cells derived from human iPS.
- Corning® Matrigel®, growth factor reduced, phenol red-free (BD, cat. no. 356231).
- Dulbecco's modified Eagle medium (DMEM, Sigma-Aldrich, St. Louis, MO, USA).
- Fetal bovine serum (FBS; e.g., Gibco, Life Technologies, Massachusetts, USA).
- Penicillin/streptomycin mixed solution (Nacalai Tesque, Japan).
- Phosphate-buffered saline (Dulbecco's formula PBS) (Genesee Scientific, El Cajon, USA).
- Potassium chloride (KCl) (Sigma-Aldrich, cat. no. P9333).
- Potassium dihydrogen phosphate (KH2PO4) (Sigma-Aldrich, cat. no. P0662).
- MEM non-essential amino acids solution (×100) (Thermo Fisher Scientific, Waltham, MA, USA).
- L-Glutamine (Nacalai Tesque, Japan).
- Sodium chloride (NaCl) (Sigma-Aldrich, cat. no. 746398).
- 2-Mercaptoethanol (Sigma-Aldrich, St. Louis, USA).
- 2.25% Trypsin-EDTA (Atlanta Biologicals, Flowery Branch, USA).
- Gelatin, tissue culture grade (type B, 2% in H_2O; Sigma-Aldrich, cat. no. G1393).
- Advanced DMEM/F12 (adDMEM/F12; Thermo Fisher Scientific, cat. no. 12634-010).
- B27 supplement, 50×, serum-free (Thermo Fisher Scientific, cat. no. 17504-044).
- BSA (Sigma-Aldrich, cat. no. 10-735-094-001).

- HEPES, 1 M (Thermo Fisher Scientific, cat. no. 15630-056).
- N-acetyl-L-cysteine (Sigma-Aldrich, cat. no. A9165).
- Nicotinamide (Sigma-Aldrich, cat. no. N0636).
- ROCK (Rho kinase) inhibitor Y-27632 dihydrochloride (Abmole Bioscience, cat. no. M1817).
- TrypLE Express Enzyme (1×), phenol red (Thermo Fisher Scientific, cat. no. 12605-010).
- Trypsin (Sigma-Aldrich, cat. no. T1426).
- 70% ethanol (Sigma-Aldrich).

20.2.2 Equipment

- Adjustable pipettes: P-20 (Gilson, cat. no. FA10003M), P-200 (Gilson, cat. no. FA10005M), and P-1,000 (Gilson, cat. no. F123602M).
- Biosafety Cabinet (BSC-04IIA2) airtech.
- Safety Cabinet Class 2 (ESCO, model no. LA2-4A1).
- Cell culture incubator (CO2 at 5%, humidified at 37°C) (Thermo Scientific cat. no. 311).
- CKX41 Inverted Microscope – Olympus Life Science (Japan).
- Cotton swabs (Sigma-Aldrich, cat. no. Z699365).
- Eppendorf 5415R Refrigerated Centrifuge (Eppendorf, Germany).
- Falcon® Conical Centrifuge Tubes (15 mL; BD Falcon, New York, NY, USA, Cat. No 352095).
- Hemocytometer (Sigma-Aldrich, cat. no. Z359629).
- Laser scanning confocal microscope, FV-1000 (Olympus, Tokyo, Japan).
- Liquid N2 storage tank.
- Microcentrifuge 1.5-mL tubes (Eppendorf, cat. no. 0030120086).
- Pipette tips 20 μL (Starlab, cat. no. S1120-1810), Pipette tips: 200 μL (Starlab, cat. no. S1120-8810), Pipette tips 1000 μL (Starlab, cat. no. S1120-1830).
- Sterile forceps.
- Water bath (Thermo Scientific cat. no. 152-4101).
- Tissue culture-treated plate, 60 mm dish (TPP Techno Plastic Products AG Schaffhausen, Switzerland, Cat. No 93060).
- Tissue culture-treated plate, 100 mm dish (TPP Techno Plastic Products AG Schaffhausen, Switzerland, Cat. No 93100).
- 5/10/25-mL plastic disposable pipette.

20.2.3 Reagents setup

20.2.3.1 N-Acetyl-L-Cysteine

Dissolve 815.95 mg of N-acetyl-L-cysteine in 10 mL of H_2O to make a 500 mM stock solution, filter and sterilize the solution. To avoid freeze-thaw cycles, make

500-μL aliquots and store at –20 °C for ≤2 months. Use this stock solution at a ratio of 1:400 (final concentration: 1.25 mM).

20.2.3.2 Nicotinamide

Dissolve 12.2 g of nicotinamide in 100 mL of PBS to make a 1 M stock solution. To avoid freeze-thaw cycles, make 5-mL aliquots and store at –20 °C for ≤2 months. Use this stock solution at a ratio of 1:100 (final concentration: 10 mM).

20.2.3.3 ROCK Inhibitor

Dissolve 50 mg of ROCK inhibitor in 1.56 mL of DMSO to make a 100 mM stock solution. To avoid freeze-thaw cycles, make 50-μL aliquots and store at –20 °C for ≤2 months. Use this stock solution at a ratio of 1:10,000 (final concentration: 10 μM).

20.2.3.4 Trypsin Solution

Dissolve trypsin to 0.125% (wt/vol) in adDMEM/F12+++ medium and store at –20 °C for ≤1 year until use.

20.2.3.5 Medium for Cancer Stem Cells Culture

For cancer stem cells, prepare high glucose containing DMEM supplemented with final concentrations of 15% FBS, 0.1 mM MEM non-essential amino acids solution, 2 mM L-Glutamine, 0.1 mM 2-mercaptoethanol, 50 U/mL penicillin/streptomycin. Store at 4 °C.

20.2.3.6 Matrigel Preparation

Thaw aliquots of Matrigel slowly at 4 °C overnight. Mix 100 μL of Matrigel and 300 μL of cold DMEM (without FBS) in a sterile 1.5-mL microcentrifuge tube.

20.2.3.7 PBS (pH 7.40) Preparation

Prepare PBS by mixing 137 mM NaCl, 10 mM Na_2HPO_4, 2.7 mM KCl, and 1.8 mM KH_2PO_4. Prepare the buffer in distilled water (dH_2O), adjust the pH to 7.4, and autoclave it for 20 min.

20.2.3.8 0.1%(w/v) Gelatin

Dissolve 0.5 g of gelatin (from porcine skin) in 500 mL distilled water and autoclave. Store at 4 °C.

20.3 Methods

Recently, 3D culture methods have been described for different cells to recapitulate growth with the features in vivo, allowing heterogeneous phenotypes including those self-renewing and differentiating to exist within the culture (Shamir and Ewald 2014; Boj et al. 2015; McCracken et al. 2014; Smith et al. 2012; Lancaster and Knoblich 2014; Lancaster et al. 2013; Gao et al. 2014; Karthaus et al. 2014; Sato et al. 2009). Here, we describe cancer organoid culture systems using human iPS-derived CSCs which developed under the inflammatory microenvironment mimicked by cancer cell lines. Briefly, human iPSCs were treated with the conditioned medium of cancer-derived cells for 4 weeks (Fig. 20.1). If the surviving cells were confirmed to have self-renewal and differentiation potential, they were impeded in Matrigel for 14 days in the presence of cancer organoid medium. The growth of the organoid could be judged for the tumorigenicity in vitro, which confirms the establishment of CSCs from iPSCs. This model of cancer organoid

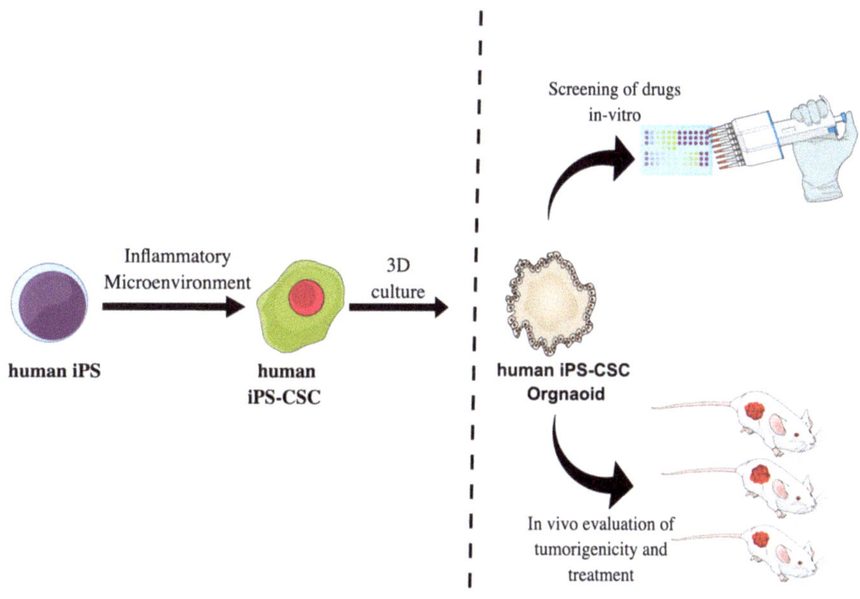

Fig. 20.1 Representative scheme to prepare cancer organoids from human iPSCs and the possible applications in the future

should feasibly be a cancer model in vitro in place of in vivo model and available for drug screening. This system will also allow the study of heterogeneous cell–cell relationships as well as the effects of different inflammatory conditions on the cells.

20.3.1 Reviving Cancer Stem Cells

1. Pick up Matrigel from a freezer at −20 °C.
2. Thaw Matrigel in a refrigerator at 4 °C overnight and then put it on ice.
3. Add 2 mL of ice-cold DMEM to a 15-mL tube on ice and add 100 uL of Matrigel with cold tips. Note: tips should be kept cold for at least 30 min in a freezer before use.
4. Transfer Matrigel in DMEM to a 60-mm dish.
5. Incubate the dish at 37 °C in 5% CO_2 for at least 1 h.
6. Warm a water bath to 37 °C.
7. Add 5 mL of Repro Stem medium to a 15-mL conical tube.
8. Pick up the vail of CSCs from a liquid nitrogen locator.
9. Thaw the vial in the water bath (Note 1).
10. Transfer the cell suspension to the conical tube prepared at Step 7 with gentle pipetting 1–2 times.
11. Centrifuge the cells at 800 rpm (160 × g), at 25 °C for 5 min.
12. Aspirate the supernatant.
13. Tap the pellet gently to break.
14. Resuspend the cells in 1 mL of Repro stem medium.
15. Pick up dishes coated with Matrigel from the incubator.
16. Add 1 mL of FF2 media to the cell suspension.
17. Aspirate excess media from the Matrigel-coated dish.
18. Wash the dish with 2 mL of ice-cold DMEM.
19. Add 4 mL of FF2 media to the Matrigel-coated dish.
20. Transfer the cells over the Matrigel-coated dishes.
21. Observe the CSC by microscopy (Fig. 20.2).

20.3.2 Sphere Formation

1. Aspirate the medium when the cells reach 80% confluence.
2. Add 1 mL of 0.05 dissociation buffer.
3. Incubate the dish at 37 °C for 5 min.
4. Stop dissociation by adding 1 mL of FF2 medium.
5. Transfer the cell suspension to a 15-mL tube.
6. Centrifuge the tube at 500 rpm.
7. Aspirate the supernatant without disturbing the pellet.

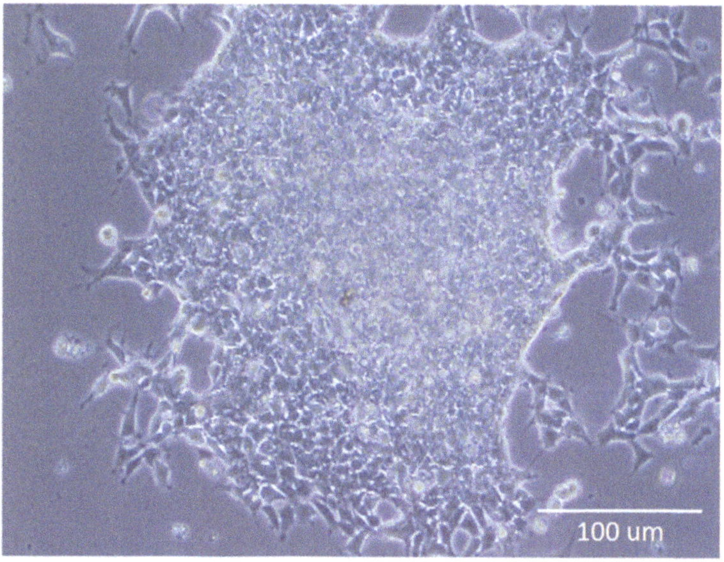

Fig. 20.2 Human iPS-CSC passage on Matrigel-coated dish

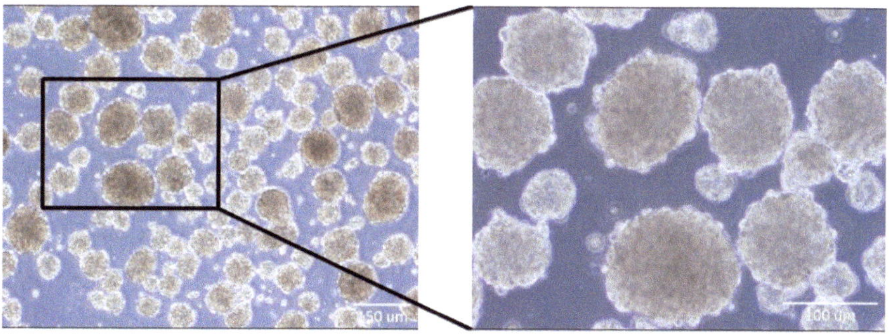

Fig. 20.3 Spheroids derived from human iPS-CSC after 10 days

8. Resuspend the cell pellet in 2 mL of FF2 and count the number (Note 2).
9. Plate 5000 cells/well in a low attachment 12-well plate to generate spheres.
10. Check the sphere every day under a microscope.
11. At approximately day 10 spheres get ready to start forming the cancer organoid (Fig. 20.3).

20.3.3 Cancer Organoid Development

1. Collect the spheres from the dish.
2. Immerse spheres into several droplets of undiluted cold Matrigel (Note 3).

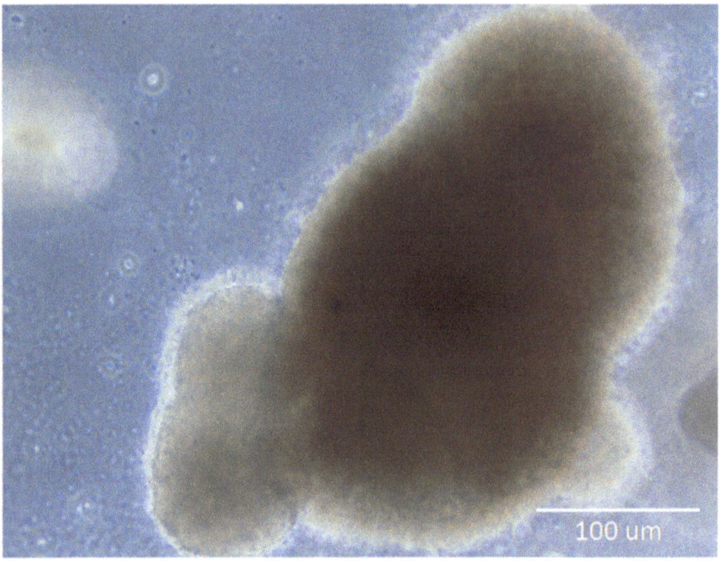

Fig. 20.4 Human iPS-CSC derived organoid after 14 days

3. Transfer the spheres in droplets to parafilm.
4. Place the sterilized parafilm with droplets at 37 °C in a 5% CO_2 humidified incubator for 1 h until the Matrigel hardens.
5. Place each droplet on a Matrigel-coated 24-well plate.
6. Place the dish at 37 °C in a 5% CO_2 humidified incubator for 1 h.
7. Add cancer organoid medium containing DMEM/F12, 1% Glutamax, 10-mM HEPES, 1:50 B27 supplement (without vitamin A), 1.25-mM N-acetyl-L-cysteine, 10-mM nicotinamide, 10 μM Y27632, and 3 nM dexamethasone supplemented with 1% penicillin/streptomycin.

Alternative Method Continuing from Step 4

8. Place the droplets in a low attachment 60-mm dish filled with cancer organoid medium.
9. Incubate the plate/dish at 37 °C in a 5% CO_2 humidified incubator for 14 days (Note 4).
10. Take photos of cancer organoids with an inverted microscope IX80 (Olympus, Japan) equipped with a digital camera device (Fig. 20.4) (Note 5).

20.3.4 Passage of the Organoids

1. Remove media paying careful attention not to lose the organoids.
2. Wash the wells with PBS twice.
3. Add 500 μL of ice-cold cell recovery solution (Falcon) per well of a 24-well plate.
4. Incubate the plate for 30 min at 4 °C.
5. Pipette up and down to disrupt Matrigel.
6. Transfer the organoid suspension to a 15-mL conical tube.
7. Place the tube on ice.
8. Add 10 mL of ice-cold DMEM/F12 medium in order to make it easy to remove Matrigel.
9. Centrifuge the tube at 200 × g for 5 min at 4 °C.
10. Aspirate the supernatant without disturbing the pellets.
11. Add 1 mL of TrypLE (Gibco) to split the organoids.
12. Incubate the tube at 37 °C until organoids fall apart.
13. Pipette up and down the suspension with a P1000-tip every 3–5 min to accelerate the disruption of cancer organoids (Note 6).
14. Centrifuge the tube at 300 × g for 5 min at 4 °C.
15. Remove the supernatant.
16. Wash the cell pellet in ice-cold PBS with gentle resuspension.
17. Centrifuge the tube at 200 × g for 5 min at 4 °C.
18. Remove the supernatant without disturbing the pellet.
19. Resuspend the pellet in Matrigel (Note 7).
20. Plate the organoids in Matrigel on the bottom of 12-well culture plates in drops of 15 μL each.
21. Place the plate into a 5% CO_2 humidified incubator at 37 °C for 30 min to let Matrigel solidify.
22. Add 500 μL of organoid medium carefully to each well after the Matrigel drops solidify (Note 8).
23. Place the culture plate in a 5% CO_2 humidified incubator at 37 °C.
24. Change the medium every 2–3 days with a fresh one by careful aspiration (Note 9).

20.3.5 Freezing of Cancer Organoids

1. Remove the medium and add 500 μL of ice cold DMEM/F12 medium into each well.
2. Break down the Matrigel drops using a P1000 filter tip.
3. Transfer the cells to an ice-cold 15-mL conical tube.
4. Fill up the 15-mL conical tubes of the organoids with ice-cold DMEM/F12 medium up to a total volume of 10 mL.
5. Pipette up and down with a Pasteur glass pipette.

6. Put the tube on ice for 10 min.
7. Pipette up and down again several times with a 10-mL serological pipette (Note 10).
8. Centrifuge the tube at 200 × g for 5 min at 4 °C.
9. Aspirate the supernatant without disturbing the pellet.
10. Resuspend the pellet in a 15-mL conical tube with 1 mL of DMEM/F12 medium.
11. Centrifuge the tube again at 200 × g for 5 min at 4 °C.
12. Aspirate the supernatant without disturbing the pellet.
13. Add the required volume of Cell-Reservoir One (Vitrify) (Nakalai Tesque).
14. Pipette up and down the organoids with a P1000 pipette to resuspend the organoids homogeneously.
15. Write the name of the samples on the vial with a water-proof pen.
16. Transfer 500 μL of each aliquot of the resuspended organoids into a cryogenic vial.
17. Place the cryogenic vial into a liquid nitrogen tank (approximately at −180 °C) (Note 11).

Notes

1. Thawing should be done within 1 min, until only small ice particles remain.
2. Cells can be kept on ice for up to 1 h before use.
3. A droplet was prepared with 10–15 μL of Matrigel on a piece of sterilized parafilm prepared in a 100-mm dish.
4. Change half of the medium every 2 days with fresh cancer organoid medium without disturbing the Matrigel droplets.
5. At day 14 the established cancer organoids are ready for further evaluation.
6. The incubation time in TrypLE should be minimum. Do not dissociate in TrypLE for more than 15 min as this may result in poor growth or loss of cancer organoids.
7. Matrigel should be kept on ice to prevent it from solidifying. Be careful not to dilute Matrigel too much with the residual PBS to form solid droplets.
8. Do not add medium directly on the top of the Matrigel drops so as not to disrupt the droplets.
9. The fresh medium should be pre-warmed.
10. The longer the sample is kept on ice, the better the Matrigel dissolves.
11. Cryopreserved organoids can be stored in the liquid nitrogen tank for several years.

References

Baker BM, Chen CS. Deconstructing the third dimension: how 3D culture microenvironments alter cellular cues. J Cell Sci. 2012;125(13):3015–24.

Boj SF, Hwang CI, Baker LA, Chio II, Engle DD, Corbo V, et al. Organoid models of human and mouse ductal pancreatic cancer. Cell. 2015;160(1-2):324–38.

Gao D, Vela I, Sboner A, Iaquinta PJ, Karthaus WR, Gopalan A, et al. Organoid cultures derived from patients with advanced prostate cancer. Cell. 2014;159(1):176–87.

Horning JL, Sahoo SK, Vijayaraghavalu S, Dimitrijevic S, Vasir JK, Jain TK, Labhasetwar V. 3-D tumor model for in vitro evaluation of anticancer drugs. Mol Pharm. 2008;5(5):849–62.

Hsiao AY, Torisawa YS, Tung YC, Sud S, Taichman RS, Pienta KJ, Takayama S. Microfluidic system for formation of PC-3 prostate cancer co-culture spheroids. Biomaterials. 2009;30(16): 3020–7.

Karthaus WR, Iaquinta PJ, Drost J, Gracanin A, van Boxtel R, Wongvipat J, et al. Identification of multipotent luminal progenitor cells in human prostate organoid cultures. Cell. 2014;159(1): 163–75.

Kievit FM, Florczyk SJ, Leung MC, Veiseh O, Park JO, Disis ML, Zhang M. Chitosan–alginate 3D scaffolds as a mimic of the glioma tumor microenvironment. Biomaterials. 2010;31(22): 5903–10.

Lancaster MA, Knoblich JA. Generation of cerebral organoids from human pluripotent stem cells. Nat Protoc. 2014;9(10):2329–40.

Lancaster MA, Renner M, Martin CA, Wenzel D, Bicknell LS, Hurles ME, et al. Cerebral organoids model human brain development and microcephaly. Nature. 2013;501(7467):373–9.

McCracken KW, Cata EM, Crawford CM, Sinagoga KL, Schumacher M, Rockich BE, et al. Modelling human development and disease in pluripotent stem-cell-derived gastric organoids. Nature. 2014;516(7531):400–4.

Sato T, Vries RG, Snippert HJ, van de Wetering M, Barker N, Stange DE, et al. Single Lgr5 stem cells build crypt-villus structures in vitro without a mesenchymal niche. Nature. 2009;459 (7244):262–5.

Seno A, Kasai T, Ikeda M, Vaidyanath A, Masuda J, Mizutani A, Murakami H, Ishikawa T, Seno M. Characterization of gene expression patterns among artificially developed cancer stem cells using spherical self-organizing map. Cancer Inform. 2016;15:163–78.

Shamir ER, Ewald AJ. Three-dimensional organotypic culture: experimental models of mammalian biology and disease. Nat Rev Mol Cell Biol. 2014;15(10):647–64.

Smith SJ, Wilson M, Ward JH, Rahman CV, Peet AC, Macarthur DC, et al. Recapitulation of tumor heterogeneity and molecular signatures in a 3D brain cancer model with decreased sensitivity to histone deacetylase inhibition. PLoS One. 2012;7(12):e52335.

Szot CS, Buchanan CF, Gatenholm P, Rylander MN, Freeman JW. Investigation of cancer cell behaviour on nanofibrous scaffolds. Mater Sci Eng C. 2011;31(1):37–42.

Yamada KM, Cukierman E. Modeling tissue morphogenesis and cancer in 3D. Cell. 2007;130(4): 601–10.

Zou Y, Cheng C, Omura-Minamisawa M, Kang Y, Hara T, Guan X, Inoue T. The suppression of hypoxia-inducible factor and vascular endothelial growth factor by siRNA does not affect the radiation sensitivity of multicellular tumor spheroids. J Radiat Res. 2010;51(1):47–55.

MIX
Papier aus verantwortungsvollen Quellen
Paper from responsible sources
FSC® C105338

If you have any concerns about our products,
you can contact us on
ProductSafety@springernature.com

In case Publisher is established outside the EU,
the EU authorized representative is:
Springer Nature Customer Service Center GmbH
Europaplatz 3, 69115 Heidelberg, Germany

Printed by Libri Plureos GmbH
in Hamburg, Germany